An Introduction
to
Scientific, Symbolic,
and
Graphical Computation

An Introduction
to
Scientific, Symbolic,
and
Graphical Computation

Eugene Fiume
University of Toronto
Toronto, Canada

A K Peters
Wellesley, Massachusetts

Editorial, Sales, and Customer Service Office

A K Peters, Ltd.
289 Linden Street
Wellesley, MA 02181

Library of Congress Cataloging-in-Publication Data

Fiume, Eugene L.
    An introduction to scientific, symbolic, and graphical computation
    / Eugene Fiume.
      p.   cm.
    Includes index.
    ISBN 1-56881-051-2
    1. Electronic data processing. I. Title.
    QA76.F54   1995
    510'.285'51--dc20   94-46591
                                            CIP

Printed in the United States of America
99 98 97   10 9 8 7 6 5 4 3 2

To Susan, Laura, and Anna

Contents

Preface

Modern computing environments support many different tools and languages. We have at our disposal powerful microcomputers, workstations, parallel processors, graphics displays, special-purpose hardware, database tools, plotting packages, symbolic computation languages, and a wide variety of numerical toolboxes. These advances have combined somewhat haphazardly. Nevertheless there is no doubt that the paradigms in scientific computation are evolving. A mathematical model can now be interactively prototyped and explored using symbolic and graphical techniques, and once confidence and understanding has been gained, the model can be transformed into an efficient and robust computational solution. Most researchers take for granted these new tools, but our undergraduate curricula have been slow to change.

This book is an introduction to mathematical representation and computation. We shall draw from a broad array of techniques from scientific, symbolic, and graphical computation. A basic course in single-variable calculus is the only real prerequisite (or corequisite), although some programming experience would be helpful. The material is intended to be accessible to undergraduates at all levels, but it is intended to appeal in different ways depending on needs and interests. It is targeted at all students in the pure and applied sciences, mathematics, and computer science who may be interested in seeing how computation interacts with a continuous world.

The choice of material for this book was (and remains) problematic: the number of topics from which to choose is huge, the literature even larger, and the mathematical foundations can be daunting even to those of us who supposedly should know better. One way to address this diversity is to choose a simple theme and follow some of its paths through various areas of mathematical computation. While we cannot cover all the territory underfoot, we may at least get a sense of the lie of the land. Accordingly, most of the material revolves around the following general questions:

(a) How can we represent a lot by a little?

(b) How can we do a lot with a little?

The first deals with the compact representation of a thing, while the second deals with its efficient computation. These two conundrums will motivate topics such as implicit and parametric representations, interpolation, approximation, sampling, filtering, reconstruction, and integration. Along the way, we shall discuss global and piecewise polynomial interpolation, splines, polynomial and Fourier series, filters, quadrature, the sampling theorem, and the solution of non-linear equations.

Because of the minimal curriculum requirements assumed by this book, namely first-year calculus, but not linear algebra, we devote very little space to problems that are best solved using general matrix formulations. It may therefore be surprising to some that we do not consider problems such as the solution to linear systems of equations (except very briefly in a simple setting), condition numbers, matrix decomposition, eigenvalues, and so on. A tempting case for an introductory book based only on linear-algebraic concepts can be made. I have also avoided the great temptation of grounding this material in one or more of the sciences, opting instead for a general core of ideas that leaves instructors free to ground the techniques to their own specialities.

There are three central beliefs to this book. The first is that graphical and symbolic exploration of concepts is crucial to understanding, and that today's computing environments provide us with sufficiently good tools to learn by doing. The second is that, largely because of the first belief, this material is suitable for first- or second-year undergraduates without significant training in computer science or mathematics, despite the fact that this book is very much about both, and that the technical content of the material we cover is of a high level. Since much of the work in elementary scientific computation and computer graphics is developed from the basic notion of the derivative as learned in high school, I believe that this material works best on students with the definitions of first-principles calculus still fresh in their minds. The success of this approach depends strongly on the availability of workstations or personal computers for students to do lab work. The third belief is that approximation theory (including, of course, interpolation) and signal theory are two sides of the same coin. I am quite mystified as to why students tend to see one topic but not the other, depending on the undergraduate programme the student has chosen to enter. A numerical analyst should understand why a CD player works, and a signal theorist should know why parametric representations are wonderful for modelling teapots!

Deciding among the tools to discuss in this book was almost as difficult as the choice of material. There are many excellent graphics systems, plotting packages, algebraic and symbolic mathematics languages, and scientific toolkits in existence. To ignore them is to ignore the current reality, but neither is it possible to discuss even the main systems. Our programming examples therefore will be illustrated using the Turing and C programming languages, and our symbolic mathematics system will be Maple. Nothing in our presentation actually depends on this choice, and it is quite straightforward to substitute another programming language such as FORTRAN, Pascal, or Modula, or to use another symbolic mathematics system such as Mathematica. Indeed, we expect that our electronic software repository (see below) will eventually support most of these languages.

We cannot be exhaustive in this book. We do not have the time to prove theorems or even be particularly rigorous. This is the domain of more advanced courses in several areas. Our main aim is to introduce some of the problems associated with converting mathematical models into computational solutions. The simple statement of the aim belies its difficulty, especially if we were to derive general solutions. On the other hand, our coverage of some topics will be surprisingly deep and will be good preparation for senior courses in scientific and mathematical computation, computer graphics, image processing, computer vision, and signals and systems.

Most of the exercises should be worked through as part of the reading process, and working through them using a symbolic mathematics system is encouraged. A few of the exercises involve a longer stay at a workstation. The ordering of the chapters beyond Chapter 2 is due more to taste than to dependencies. Potential instructors may be surprised to see that floating-point representations are not formally presented until Chapter 5. If students have had prior exposure to series representations from calculus, then this material could be presented earlier, although it is not essential to do so.

Acknowledgements

I am grateful to Wayne Enright, Sherif Ghali, Michael McCool, and Chris Williams, for reading, criticising and improving this book. Earlier drafts of this book were inflicted on the students of CSC160 and CSC260 at the University of Toronto, and many students took the time offer suggestions and corrections. I thank them for their help and patience. Allan Borodin encouraged me to design a course that got this whole thing started; blame me for the implementation. Alain Fournier has inspired me in many ways, but in this book he has specifically inspired my treatment of polynomials and parameterisation in Chapters 1 and 2. The gang at AK Peters, namely Alice and Klaus Peters, Alexandra Benis and Joni McDonald have been a treat to work with.

Suggestions for Further Reading

The main reason for the existence of this book is that there are few, if any, books on mathematical computation that are directed at the beginning undergraduate. However, a great many useful, excellent, but more specialised books exist at a higher level. The list below is certainly not exhaustive, and is largely an indication of my influences and tastes.

Advanced Calculus and Real Analysis

These are both excellent introductions to real analysis. The first has a particularly attractive discussion of Fourier series.

J.E. Marsden, *Elementary Classical Analysis*, W.H. Freeman and Co., San Francisco, 1974.

W. Rudin., *Principles of Mathematical Analysis*, Second Edition, McGraw-Hill, New York, 1964.

Scientific Computation

The first is a good introduction to many of the issues in scientific computation, while the second is thorough treatment of quadrature, including Monte Carlo integration.

D. Kahaner, C.M. Moler, and S. Nash, *Numerical Methods and Software*, Prentice-Hall, New York, 1989.

P. Davis and P. Rabinowitz, *Methods of Numerical Integration*, Second Edition, Academic Press, New York, 1984.

Approximation Using Polynomials

The first two books are standard texts on polynomial curves and surfaces as relevant to computer-aided geometric design and computer graphics. The third is a comprehensive and remarkably well written presentation of similar material that is my current favourite. The last reference is an advanced but accessible presentation of approximation methods for many problems in scientific computation.

R. Bartels, J. Beatty and B. Barsky, *An Introduction to Splines for Use in Computer Graphics and Geometric Modeling*, Morgan Kaufmann, Los Altos, CA, 1987.

G. Farin, *Curves and Surfaces for Computer-Aided Geometric Design* Third Edition, Academic Press, Boston, 1993.

J. Hoschek and D. Lasser (translated by L.L. Schumaker), *Fundamentals of Computer-Aided Geometric Design*, A.K. Peters, Wellesley, MA, 1993.

P. Prenter, *Splines and Variational Methods*, Wiley, New York, 1975.

Signal Theory and Processing

The first book is an authoritative introduction to the theory and processing of general signals. The second is an advanced treatment of digital image processing. Both are wonderful books.

A.V. Oppenheim, A.S. Willsky, and I.T. Young, *Signals and Systems*, Prentice-Hall, Englewood Cliffs, NJ, 1983.

W. Pratt, *Digital Image Processing*, Second Edition, John Wiley and Sons, New York, 1990.

Language References

Useful books containing illustrative examples of the use of symbolic programming languages are appearing almost daily. While Maple is used extensively in this book, the material does not depend on this choice for a symbolic computation language. Nevertheless, two references are included in this list. Note that the book by Burbulla and Dodson has an equivalent version for Mathematica. Introductions to Mathematica and to MATLAB are also listed.

N. Blachman, *Mathematica: A Practical Approach*, Prentice-Hall, Englewood Cliffs, NJ, 1992.

D.C.M. Burbulla and C.T.J. Dodson, *Self-Tutor for Computer Calculus using Maple*, Prentice-Hall, Englewood Cliffs, NJ, 1993.

B. Char *et al.*, *Maple V Language Reference Manual*, Springer-Verlag, New York, 1992.

The Math Works, Inc., *The Student Edition of MATLAB*, Prentice-Hall, Englewood-Cliffs, NJ, 1992.

Internet Connectivity, or How You Can Make a Difference

One of the remarkable outgrowths of the Internet is the increasing availability of electronic mechanisms for information dissemination. A repository of exercises, sample software, and applications exists. A separate repository of solutions is also available for instructors, as is an initial distribution of software for courseware. Instructors and students alike are welcome to contribute to the repository. An information page can be found via the World-Wide Web at

$$\text{http://www.dgp.toronto.edu/~elf}$$

A synonym for dgp.toronto.edu is dgp.utoronto.ca.

Footnote to Second Printing

The short time since the first printing has seen an blossoming of more tools for mathematical computation. The most notable of these are a greatly extended MatLab, and the ubiquitous Java. My web page now points to a variety of new resources for mathematical computation. In this printing, I have restricted myself to correcting errors, most of which were typographical; some, however, were more significant (i.e., embarrassing). I am thankful to John Hart and James Stewart for their suggestions. I am especially grateful to Radford Neal for the dozens of useful comments that he provided.

Eugene Fiume
May, 1997

Chapter 0
Mathematical Computation

As we stated earlier, this book is about representing a lot by a little, and about doing a lot with a little. How is it that so much music can be represented on a small compact disk? How can the trajectory of an object be modelled using a small number of values, and why can it be depicted so quickly? Why is so little information needed to generate a fractal or to represent a surface? How can the details of a three-dimensional brain scan be made so vivid when the scanning devices are so noisy? We will be able to answer questions of this type as we progress through this book, but we need to understand how to put fundamental mathematical concepts to work for us as well. Our discussion will touch on many overlapping areas of study. The relationship among some of them is discussed in this chapter.

0.1. Scientific, Symbolic, and Graphical Computation

The richness of our science, technology, and art long ago escaped complete comprehension by any single person. For better or worse, our collective attempt to understand the world is performed by using a divide-and-conquer strategy: concepts and problems are parcelled into chunks that have some hope of being understood by specialists such as scientists, mathematicians, engineers, philosophers, and artists. Each area of specialisation has evolved its own methodologies and techniques that best fit the structure of their problems. In reality, the "chunks" are not particularly well-defined; the problems and preoccupations of these specialities overlap on the one hand, but there may be gaps among them on the other. Worse still, our ignorance may be so great that we may not even properly identify or assign a chunk to an appropriate speciality.

1

Somewhere among these specialisations sits computer science. Our discipline is itself broadly based in the sciences, humanities, engineering and mathematics. We are preoccupied with the structure of problems and their computational solution. We view the solution to large problems, both abstractly and concretely, as a system of small solutions that co-operate through well-defined interfaces. The qualities we hold in high regard include correctness, efficiency, reusability, robustness, and "elegance." We are among the engineers when we construct large software systems. We are among the mathematicians when we prove properties of our systems. Finally, if the behaviour of our solutions resists formal proof, we may resort to experiments to validate and to understand more fully their computational behaviour. Computer science would be a less vigorous field if, in forging itself as a separate discipline, it did not borrow methodology from its relatives.

Computers and computation have affected almost every facet of human activity. Our goal in this book is to look at one facet of activity: the interplay between the mathematical modelling and computation. Traditionally, the branch of computer science (and mathematics) most concerned with this activity is called *numerical analysis*. This discipline is concerned with methodologies for both developing, and evaluating the quality of, implementations of mathematical models on calculating devices. Numerical analysis has its roots in Newton, Leibnitz, Euler, and Gauss, who made fundamental contributions to the numerical solution of problems in calculus and linear algebra. The first computational numerical analyst was probably Charles Babbage, who in the 19th Century proposed a remarkable steam-powered machine to compute "forward differences." He called the machine the *difference engine*. To Babbage's deep disappointment, only a small portion of it was built in his lifetime.[1] We shall have more to say about forward differences later, but Babbage clearly appreciated a fundamental tenet of numerical analysis: that computational implementations are fraught with delicate issues of representation and discretisation, and that the accuracy of computational solutions must be precisely characterised with respect to these issues. The choice, for example, of a specific representation for real numbers, together with a set of operations on the representation, must be well-understood if we are to have confidence with the accuracy of a solution involving this representation.

Modern numerical analysis is closely allied with *scientific computation*. While the nomenclature is still debated among researchers in the field, this term is generally used to denote a huge area of research and application involving the numerical solution of problems in the physical sciences. The solutions to these problems typically involve the numerical solution of linear, nonlinear, or differential equations, numerical optimisation, interpolation, and integration.

---

1. An entertaining speculative novel that explores what might have happened if only the difference engine had been completed is *The Difference Engine* by William Gibson and Bruce Sterling.

These are problems defined over a *continuous* domain such as the real numbers. The body of *numerical methods* to implement computer solutions to these problems is staggering. Introductions to these methods are found in senior undergraduate computer science or applied mathematics courses. Because numerical methods require that ideal "real" numbers have an *explicit* numerical representation, and because these representations and operations on them are usually imprecise, the characterisation of the *numerical error* made in a calculation is an important preoccupation. Hence the continuing relationship between numerical analysis and scientific computation. There are, however, problems in scientific computation that are not within the domain of numerical analysis. For example, the implementation of algorithms on parallel machines or in hardware may have the same numerical considerations as those implemented on conventional computers, but may have a much different methodology associated with their implementation. In this case, the numerical analysis is the same, but there are different computational issues to be considered.

In the meantime, a body of tools and approaches has emerged that take, at the outset at least, a different view of solving mathematical and scientific problems. Another focus of mathematical computation has been to employ *implicit* or *symbolic* representations of ideal mathematical objects. These representations are particularly convenient when working with problems involving discrete forms of mathematics (which is not to say that they lack a continuous component): algebraic geometry and topology, logic, graph theory, number theory, and polynomials. Because of the nature of the representations, this area is also called *symbolic computation*. One of the topics of this book is the application of symbolic and scientific computation to a variety of problems that can be solved using both strategies.

For lack of a better term, we shall use *graphical computation* to mean computations involving visual representations. As our computing technologies have evolved, we have placed increasing reliance on the graphical presentation of data. As we shall see, the ability to explore a phenomenon graphically can convey essential aspects of its behaviour. Computer graphics is therefore assuming a significant role both in and out of scientific computation. Conversely, the needs of the sciences are influencing computer graphics. Long concerned with the modelling and rendering of phenomena using definite geometric primitives such as polygons, polyhedra, and curved surfaces, computer graphics has begun to incorporate techniques for the rendering of unstructured volumetric data. In many cases, the output from the medical and scientific applications is a volume of data samples indicating measurable quantities such as the density or opacity of a material at a position in space. For example, the output of a magnetic resonance imaging (MRI) system is a three-dimensional dataset with a given point in the data indicating the relative density of the material at the corresponding point in the object. Although the physical quantity measured in magnetic resonance is a complex phenomenon, the resulting density values can be used to define an imaging model in which low densities can be interpreted as being more transparent than high densities (or vice versa). The resulting image is one in which a human observer may detect surfaces and other features visually, despite the fact that there may be no explicit classification of these features in the data.

We are thus distinguishing scientific, symbolic, and graphical computation by the kinds of objects each manipulates. This is certainly an oversimplification, in that the various representations and computations overlap, but these distinctions are useful. We shall broadly group these computational methods together and label them *mathematical computation*.

0.2. Themes of this Book

The emergence of powerful workstations and programming environments has greatly changed the way we both perform research on, and construct, mathematical software. We shall demonstrate how today's tools can be used in concert to implement solutions, or to evaluate solutions when more than one exist. In particular, we shall consider the effect of using symbolic and scientific computation and computer graphics co-operatively to devise good solutions to mathematical problems. There are now many systems that support these forms of computation. Their use on workstations is now entirely commonplace in research. The same tools are now available to the general public, and our teaching methods can exploit this. The following themes will arise repeatedly in this book:

- the use of symbolic computation to prototype the behaviour of models.
- the use of computer graphics to visualise results.
- the use of numerical methods to derive efficient and robust implementations after prototyping.

We shall be economical in our choice of applications: rather than consider many widely-varying problems, our problems will be linked together, and will all somehow involve the discrete representation of continuous phenomena. This is just a long-winded way of saying what we have said twice already: that we wish to represent a lot by a little. Such representations are fundamental to making feasible the use of digital computers to solve scientific problems. The very existence of much of the technology we encounter daily depends on it: most of the music to which we listen is somehow encoded using a set of digital samples represented as pits on the surface of a compact disk and then reconstructed into music; text, lines, curves, and surfaces are drawn swiftly on our computer screens; images on television or from space appear crisp and well-defined despite being engulfed by a great deal of noise; medical scanning techniques can make pictures of, or accurately monitor the condition of, our bodies.

Thus we see that there are many "graphic" examples of the importance of having good discrete representations. As a result of these concerns, we will consider at length what might at first seem to be an unlikely mix of problems in parameterisation, interpolation, integration, signal processing, and computer graphics. The inherent relationships among these topics have been insufficiently stressed in our undergraduate studies, because these topics independently arise in more advanced courses, typically in different curricula taught in a diverse set of university departments. Consequently, many of us are not exposed to the sampling theorem: a great fundamental truth, one that is so important to computation, is rarely seen outside an electrical engineering department. Similarly, we shall recognise the importance of piecewise parametric polynomials to solve interpolation and integration problems.

To summarise, this book deals with mathematical computation, which consists of both scientific and symbolic computation. We shall combine these two modes together with computer graphics to study techniques in the discrete representation of continuous phenomena. The main tools of our work will be introduced naturally in the course of our discussion. Because symbolic computation is somewhat of a paradigm shift for traditionally schooled computer scientists, we shall now give a brief example of the approach.

0.3. Symbolic Computation

The representation of mathematical objects in a symbolic rather than numeric computational form has existed since the early days of computer science. Indeed, many numerical methods are based on symbolic forms. For example, the definite integral of a polynomial

$$p(t) \;=\; \sum_{i=0}^{n} a_i\, t^i$$

on an interval $[t_0, t_1]$ with real coefficients a_i is never computed using numerical integration (also called *quadrature*). Instead, the indefinite integral of p has a closed-form solution, so that the definite integral can be evaluated directly. Similarly, the compound integration rules that we shall discuss later have, to a large extent, a symbolic foundation. The notion of "symbolic computation" is thus not new to numerical methods. The 1970s and 1980s have seen the development, however, of environments that place a greater emphasis on computation with mathematical objects in an implicit or symbolic form. Symbolic computation is based on defining objects not as numerical quantities, but as entities that have certain mathematical properties. For example, among the many properties of π, we know that it has a non-terminating, non-repeating numeric representation. We also know, on the other hand, some fundamental non-numeric mathematical relationships in which π is involved. For example:

$$\pi \;=\; \int_{-\infty}^{+\infty} \frac{\sin x}{x}\, dx.$$

Furthermore, *Leibnitz' formula* states

$$\frac{\pi}{4} = 1 - \frac{1}{3} + \frac{1}{5} - \frac{1}{7} + \cdots,$$

and *Euler's formula* is

$$\frac{\pi^2}{6} = \sum_{n=1}^{\infty} \frac{1}{n^2}.$$

A great many such relationships exist. We also know the algebraic effect of applying certain functions on π:

$$\sin \pi = 0,$$

$$e^{i\pi} + 1 = 0.$$

An environment supporting symbolic computation facilitates the manipulation and combination of these relationships. The output of a symbolic computation is often another symbolic quantity such as a series or other mathematical object that is *unevaluated*, in the sense that a numeric representation has not been explicitly computed. The ability to defer numeric evaluation and concentrate on symbolic manipulation distinguishes these computing environments from traditional numerical approaches. At some point, an expression may be evaluated, yielding a numerical quantity, but this is neither always necessary nor desirable. Two well-known and powerful symbolic computation environments are Maple and Mathematica; for our purposes they are functionally equivalent.

We have stressed the word *environment* over *language* because for most users of these systems, it is the former that they see. For example, Maple and Mathematica both support a symbolic programming language with procedures, functions, recursion, etc., but most users of these systems take advantage of simpler interactive command-line scripts as well as mouse-based dialogues that manipulate expressions and perhaps plot the results. Some of this work may later go into procedures and libraries. Every time a new relationship for π is created, it may interact with other relationships, and may provide new ways of deriving solutions to other problems. These environments permit varied, extensible, interactive use, particularly when tied to a fast graphics-capable workstation, allowing users to collect tools in an ever-growing mathematical toolchest.

0.3.1. An Example

Suppose we wish to study the convergence of Leibnitz' formula for π. Let

$$P(n) = 1 - \frac{1}{3} + \frac{1}{5} - \frac{1}{7} + \cdots + \frac{(-1)^n}{2n+1}$$

$$= \sum_{i=0}^{n} \frac{(-1)^i}{2i+1}.$$

A mathematician might prefer to write $P(n)$ as P_n, since it can be thought of as the nth term in a sequence of values converging to $\pi/4$. We write it as $P(n)$ because we shall instead think of it as a computation by a procedure P with parameter n. A Maple procedure to compute this partial series is[2]

```
P := proc (n)
     local i;
     sum( (-1)^i / (2*i+1), i=0..n );
end:
```

We can now explore the behaviour of the series. In the following, lines beginning with "`>`" are typed by the user, while the results returned by Maple are centred.

```
> P(0);
```
$$1$$

```
> P(1);
```
$$2/3$$

```
> P(10);
```
$$\frac{11757173}{14549535}$$

Notice that the assigned result is in a symbolic fractional form. To evaluate the result numerically, we type

```
> evalf(");
```
$$.8080789524$$

where the `"` refers to the previous result. A numeric approximation for $\pi/4$ is

```
> evalf(Pi/4);
```
$$.7853981635$$

which differs considerably from our approximation. However, observe that

```
> evalf(P(1000));
```
$$.7856479136$$

So, 1000 terms in the series only buys us three decimal places of accuracy. To see just how bad the convergence is, let us plot the behaviour of `P(n)` as `n` increases. We first create a sequence of values $(0,P(0))$, $(1,P(1))$, $\cdots$, $(n,P(n))$ and then plot the sequence as a connected curve, as follows.

---

2. All interactions with Maple will be printed in a **typewriter** (courier bold) font.

```
> pts := [ [i, P(i)] $ i=0..100]:
```

```
> plot(pts);
```

We used a colon above to inform Maple that we do not wish the resulting value of **pts** to be displayed. The plot appears in Figure 0.1. Notice how the alternation in the series causes oscillation about the correct result, and that convergence is exceedingly slow. On the assumption that we know the value to which the series will converge, one could proceed by defining an *error* function based on the *residual* error given by:

$$\varepsilon_n \;=\; \frac{\pi}{4} - P(n).$$

Thus ε_n tells us the signed difference between the limit of P and $P(n)$ for any n. In Maple, we could compute a sequence of such errors as follows:

```
> eps := [ [i, Pi/4 - P(i)] $ i=0..100]:
```

The resulting plot is depicted in Figure 0.2.

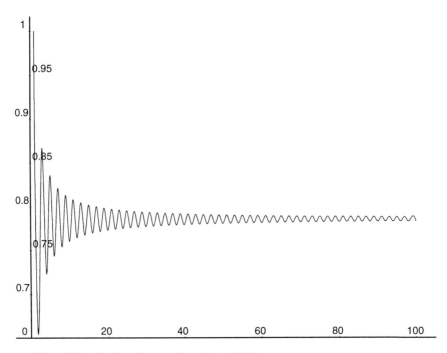

Figure 0.1.   A plot of **P(0)** through **P(100)**.

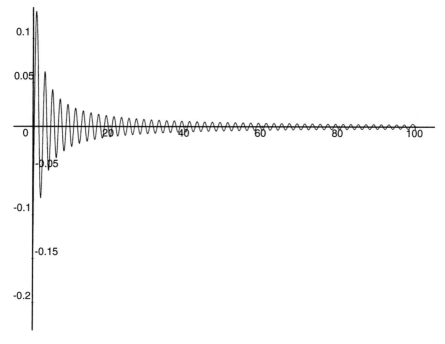

Figure 0.2. A plot of 100 terms of the residual error ε_n.

Exercise 0.1. The function $P(n)$ and various error functions are all plotted using continuous curves. Is this misleading? Discuss.

We are often interested in the magnitude of the error and not its sign. In other words, whether $P(n)$ wobbles below or above $\pi/4$ may be less important to us than the amount of deviation. The *absolute* error made by $P(n)$ is defined as

$$\alpha_n = \left| \frac{\pi}{4} - P(n) \right|.$$

In Maple, we can write this as

```
> alpha := [ [i, abs(Pi/4 - P(i))] $ i=0..100]:
```

The plot of the absolute error made by $P(n)$ in Figure 0.3 gives us additional intuition to the previous error plot.

Another common error function is based on the *proportion* of error made. Specifically, the error between the approximation and the actual value is stated relative to the actual value:

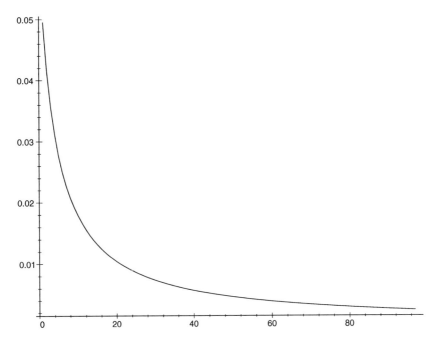

Figure 0.3.   A plot of 100 terms of the absolute error α_n.

$$\rho_n = \frac{|\pi/4 - P(n)|}{\pi/4}$$

$$= \frac{\alpha_n}{\pi/4}.$$

The *relative* error function ρ is plotted in Figure 0.4. Notice that each error function tends toward zero. Numerical analysts are concerned with the *rate* at which the error goes to zero (with respect to terms in a series or iterations of a numerical technique). All other things being equal, approximations with errors that converge more quickly to zero are considered superior. We shall have more to say about error later on in this book.

Exercise 0.2. What could be meant by the statement "all other things being equal"? How else could different approximations to the same function or value actually be different?

It is worth reviewing our philosophy. What new techniques have we been using to explore our numerical approximation for π? First, we have been using an *interactive* environment that allows us the flexibility to explore new ideas and modify them quickly. Second, graphical information display plays an important role. Third, symbolic representations allow us to explore concepts more directly.

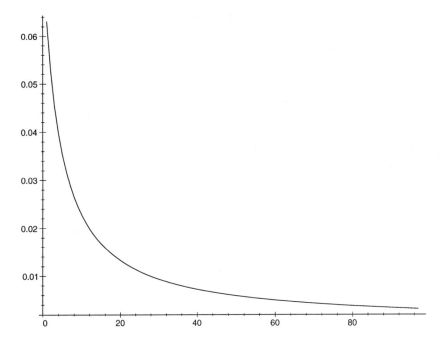

Figure 0.4.   A plot of 100 terms of the relative error ρ_n.

The use of symbolic computation is not a magic solution to our problems. For example, we have only *qualitative* evidence that if our goal is to compute π efficiently, then Leibnitz' formula may not be directly practical. We do not have a proof of the convergence rate, nor will an efficient technique be automatically found for us.

As we progress through this book, we shall see that symbolic and numeric computation are mutually supportive. Despite being quite powerful for some classes of computation, symbolic languages have their own idiosyncrasies and frustrations, and they can execute slowly by the standards of numerical methods implemented using traditional compiled programming languages. However, different purposes are served by numeric and symbolic approaches.

In computer science, we measure the *efficiency* of an algorithm or an implementation by proving properties of its running time and memory usage. A good implementation of any algorithm, and especially a numerical method, is subject to mathematical scrutiny before its implementation, and careful empirical testing afterward. When performing interactive mathematical computations, notions of space and time efficiency usage are fuzzier, because the way in which the system performs many of its functions is less clear than in a conventional programming language. There is no guarantee that a certain statement will execute quickly, but in the initial stages of our enquiry, we may be satisfied to wait a second or two if the concepts can themselves be economically expressed. Interactive programming environments may initially strike a different balance between computational efficiency and human productivity. Once we have found a potentially suitable approach, our job of making it efficient and robust begins.

Sometimes, however, computational efficiency also aids human productivity. For example, the Maple code above for the computation of various sequences was intended to be clear, but they are terribly inefficient. On basic workstations, the user may have to wait one minute or more for one of the above plots to be computed, when in fact slightly more carefully written Maple code would reduce the computation to interactive speed.

Exercise 0.3. Recall the following Maple statement that was used above.

```
> pts := [ [i, P(i)] $ i=0..n]:
```

Why is this statement inefficient? Of what order is it (in n)? Write a Maple procedure that is many times faster than this. Your routine should be $O(n)$. The notation $O(f(n))$ for a function $f(n)$ means that the running time (or space) of the algorithm is bounded above by $kf(n)$ for some constant $k > 0$. The value n normally refers to the *size* of the input, such as the number of elements, length of a sequence, etc.

Exercise 0.4. Using the plotting techniques and error functions described above, examine graphically the convergence behaviour of *Euler's formula*:

$$\frac{\pi^2}{6} = \sum_{n=1}^{\infty} \frac{1}{n^2}.$$

Exercise 0.5. We know that the slope or tangent of a 45-degree line is 1. That is,

$$\tan \frac{\pi}{4} = 1.$$

We can invert the equation by applying an inverse tangent to both sides:

$$\tan^{-1}(\tan \frac{\pi}{4}) = \tan^{-1} 1,$$

so that

$$\frac{\pi}{4} = \tan^{-1} 1. \tag{1}$$

The inverse tangent is called **arctan** in Maple. From Chapter 5 (and from calculus), we know that many functions can be approximated by a series expansion. The **series(f,x,n)** operation in Maple computes n terms of a Taylor series for the function $f(x)$. Use Maple and Eq. 1 to to deduce Leibnitz' formula. (This problem is easy to answer even if you don't know what a Taylor series is. Just use Maple!)

0.3.2. A More Complex Example

The following example derives a more subtle algorithm for approximating the value of π. We saw from the previous exercise that Leibnitz' rule can be derived from a series expansion and that with successive terms in the series we alternately overestimate and then underestimate the actual value of π. We also saw that it converges slowly to the desired value. Let us construct an approximation for π that is *always* an underestimate and that, under a big assumption, converges much more quickly.

Consider a circle of diameter 1. Then its radius is ½. As we shall see in more detail in the next chapter, such a circle can be conveniently defined *parametrically*, which means that we can describe the behaviour of the circle independently in its co-ordinates, say x and y. In particular if we plot $(x(\theta), y(\theta))$, where

$$x(\theta) = \text{½} \cos \theta,$$

$$y(\theta) = \text{½} \sin \theta,$$

and $\theta \in [0, 2\pi]$, then a circle is the result. In Maple, we can (verbosely) write:

```
x := proc(t)
    1/2*cos(t);
end:


y := proc(t)
    1/2*sin(t);
end:


circle := proc()
    plot([x(t),y(t), t=0..2*Pi]);
end:
```

Upon invoking **circle()**, we get the plot of Figure 0.5.

The *circumference* of (i.e., the distance around) a circle of radius $r = $ ½ is $2\pi r$, or simply π. Thus if we can approximate the circumference we have an approximation to π. One way to do this is to compute the *perimeter* of a polygon whose edges lie close to the circle. Let us *inscribe* the circle with a polygon, meaning that the polygon lies completely within the interior of the circle except at the polygon's vertices. For simplicity, we shall assume the polygon is *regular*, meaning that its edges are all the same length. Thus if the polygon has n sides, then its perimeter is just n times the length of one side. It is very easy to compute regular polygons that are inscribed by a circle: if we are to have n edges, then we compute n equally spaced points on the circle. To do this we just compute $x(\theta)$ and $y(\theta)$ for each angle

$$\theta = i\frac{2\pi}{n}, \quad i = 1, 2, \cdots, n.$$

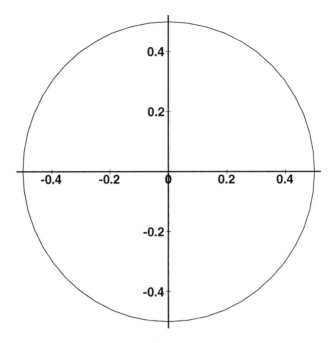

Figure 0.5.   A circle of unit diameter.

The following code plots both a circle and an inscribing regular *n*-gon.

```
ngon := proc(n)
    local i,pts;
    pts := [[x(i*2*Pi/n),y(i*2*Pi/n)] $ i=0..n];
    plot({[x(t),y(t), t=0..2*Pi],pts},style=LINE);
end:
```

Here, **x** and **y** are Maple functions as above. Figure 0.6 (left) is the result of
ngon(4) and Figure 0.6 (right) depicts **ngon(12)**. Clearly the approximation
gets to within visual tolerance of the circle quite quickly. There are many ways to
compute the length of one side of a regular polygon. The next exercise works
through one way.

Exercise 0.6. Let θ_n be the angle measured from the centre of the circle between two
endpoints of any edge of a regular *n*-gon inscribing the unit-diameter circle. For example,
for an inscribing square, $\theta_4 = \pi/2$ (as in Figure 0.7); in general, $\theta_n = 2\pi/n$, for a regular,
inscribing polygon on *n* vertices. Show that the length d_n of the edge of a regular *n*-gon is

$$d_n = \sqrt{(1 - \cos\theta_n)/2}.$$

What is d_4? What is the corresponding perimeter of the inscribing square?

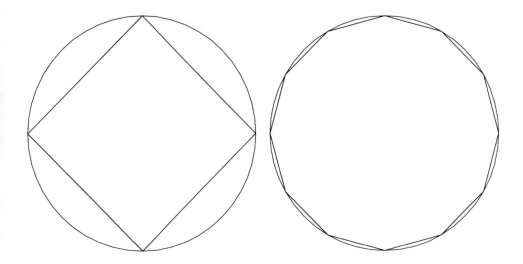

Figure 0.6. Regular polygons inscribing a circle. Left: a square. Right: a dode-
cagon.

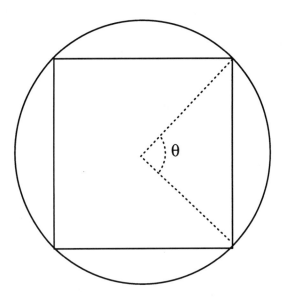

Figure 0.7. Interior angle made by an edge of an inscribing polygon.

From this exercise, the perimeter P_n is

$$P_n = n\,d_n = n\sqrt{\left(1 - \cos\frac{2\pi}{n}\right)/2} \ .$$

We can implement this formula using Maple as follows.

```
Pn := proc (n)          # perimeter of regular n-gon
     n * sqrt( (1-cos(2*Pi/n))/2 );
end:
```

When we plot the absolute error of this new formula against Leibnitz' formula as in Figure 0.8, we see that it clearly converges more quickly, at the expense of many square root and cosine computations, which are in general more expensive than simple divisions.

   We have another embarrassing problem to overcome: we are using the value of π to compute it! It is one thing to exploit the properties of π, but it is quite another to assume we have a value before we have computed it. Let us use the properties of π to modify our algorithm. Rather than computing P_n for arbitrary n, let us insist that n be a power of two. That is $n = 2^k$, for some $k > 1$.

   Since so many n are now forbidden, let us instead write Q_k for the perimeter of a regular 2^k-gon inscribing the unit-diameter circle. In other words, $Q_k = P_{2^k}$. Similarly, let ω_k be the angle between two endpoints of an edge in the 2^k-gon (so $\omega_k = \theta_{2^k}$). Thus $Q_2 = P_4 = 2\sqrt{2}$ is perimeter of the inscribing square. Observe that

$$\omega_k = 2\omega_{k+1},$$

meaning that the angles between endpoints are halved. The double-angle formula for the cosine function is

$$\cos 2\alpha = 2\cos^2 \alpha - 1.$$

This allows us to write out a recurrence. Define $C_k = \cos \omega_k$. Then

$$C_k = \cos \omega_k = \cos\frac{2\pi}{2^k} = \cos\frac{\pi}{2^{k-1}}.$$

On the other hand,

$$C_{k-1} = \cos\frac{\pi}{2^{k-2}} = \cos\left(2 \cdot \frac{\pi}{2^{k-1}}\right).$$

Applying the double angle formula on the right-hand side implies that

$$C_{k-1} = 2C_k^2 - 1,$$

meaning that

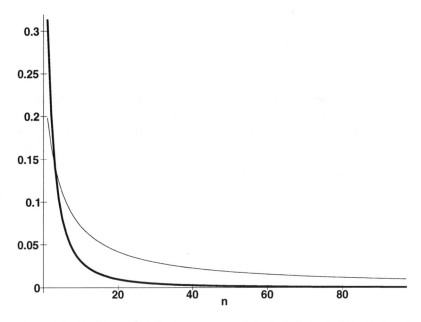

Figure 0.8. Comparison of absolute errors made by Leibnitz' rule (thin line) and perimeter computation (thick line).

$$C_k = \sqrt{(C_{k-1}+1)/2} \ .$$

This allows us to write an expression for the cosine of a small angle in terms of the cosine of an angle twice its size and a square root. We "bottom out" with C_2, which is simply $\cos \frac{\pi}{2} = 0$. Recalling d_n from Exercise 0.6, our expression for the perimeter of a 2^k-gon is

$$Q_k = 2^k \sqrt{(1-C_k)/2} \ . \tag{2}$$

Observe that we have avoided using the value of π at the expense of a recursive formula that creates a sequence of nested square roots. The Maple code to perform this computation is

```
perim := proc (k)
     2^k*sqrt((1-C(k))/2);            # recurrence avoids Pi
end:
```

```
C := proc (k)
   option remember;
   if (k=2) then 0
   else sqrt( (C(k-1)+1)/2 );
   fi;
end:
```

The **option remember** statement causes Maple to recall the result of previous computations of **C(k)**, which saves a great deal of time if a function is often invoked with the same values for arguments. Our derivation with respect to side length of a regular polygon is similar to that made by Archimedes more than 22 centuries ago. His recurrence is based directly on perimeter Q_k:

$$Q_{k+1} = 2^k \sqrt{2\left(1 - \sqrt{1 - (Q_k/2^k)^2}\right)} \ . \tag{3}$$

Kahaner *et al.* present an excellent sequence of exercises illustrating the pitfalls of numerically implementing this rule.[3] It turns out the same is true of our rule (see the exercises). However they demonstrate how this formula can be successfully refined into an efficient, robust procedure. Press *et al.* describe how one might compute the value of π to very high precision (e.g., several thousand digits).[4]

Exercise 0.7. We have seen that a plot provides a visual evidence of the rate at which an approximation converges. What would be a good quantitative measure of the convergence rate? What if the value to which a series or sequence converges is not known?

Exercise 0.8. Using Maple show that Archimedes rule, Eq. 3, agrees with our rule, Eq. 2. Test your results by subtracting one rule from the other. In two separate small procedures, use **simplify** in one and **evalf** in the other to see if the result for arbitrary k is always zero. Are there any differences between the two procedures?

Exercise 0.9. Type in the Maple code **Pn**, corresponding to P_n on page 16. For small values of n, use Maple to show that

$$P_{2^k} = Q_k.$$

Once again, use two different procedures: one using **simplify**, and the other using **evalf**. Something strange happens when using **evalf**. Investigate this behaviour. The Maple variable **Digits** allows you to control the number of digits a numerical representation has. How is the computation affected by changing **Digits**?

---

3. D. Kahaner, C.M. Moler, and S. Nash, *Numerical Methods and Software*, Prentice-Hall, New York, 1989.
4. W. Press, S. Teukolsky, W. Vetterling, and B. Flannery, *Numerical Recipes in C*, Second Edition, Cambridge University Press, Cambridge, UK, 1992.

Chapter 1
The Representation of Functions

In this chapter, we shall briefly review the familiar number systems with which we shall be working. We then discuss the common forms of functions defined over these number systems. Several representations for functions will be introduced, including explicit, implicit, parametric, and procedural forms. The parametric representation will be particularly useful to our work. We shall consider parametric functions in general, and then focus on an important subclass of parametric functions: the polynomials. The fast and accurate evaluation of functions is crucial to practical implementations. As examples of evaluation techniques, we derive several rendering algorithms for lines and circles. These examples will point out the some of the problems involved in converting mathematical representations to discrete representations such as visualisations on a display screen. Many other instances of these problems will arise in other chapters. Sections 1.1 to 1.3 are review material necessary for our discussion. Consult a calculus book for greater detail and generality.

1.1. Sets and Number Systems

Mathematicians have developed a wide variety of number systems and operations over them. For our purposes, we are interested in just a few, although there are many others, such as the complex numbers and the quaternions, that are also extremely useful.

The set of *natural numbers*, denoted **N**, is the set of non-negative whole numbers:

$$\mathbf{N} \;=\; \{0, 1, 2, \,\cdots\, \}.$$

The set of *integers*, **Z**, contains the signed whole numbers including zero:

$$\mathbf{Z} = \{\cdots, -2, -1, 0, 1, 2, \cdots\}.$$

The set of *rationals*, **Q**, also include numbers with fractions:

$$\mathbf{Q} = \{p/q : p, q \in \mathbf{Z}, q \neq 0\}.$$

The set of *real numbers*, **R**, is a somewhat more difficult set of numbers to define. Each real number can be characterised in terms of a converging sequence of rational numbers. More specifically, any element $x \in \mathbf{R}$ can be defined as the limit of a sequence of rational numbers $x_0, x_1, \cdots$ that converges to x. An irrational real number such as $\sqrt{2}$ or π can thus be thought of as being somehow constructed to arbitrary precision using operations on rational numbers. In the last chapter, for example, the sequence $P(n)$ of rational numbers converged to $\pi/4$ as $n \to \infty$. In computer systems, real numbers are usually represented explicitly by *floating-point numbers* or implicitly using a symbolic or rule-based representation. A floating-point representation must be used with caution, since a single floating-point number can represent an uncountably infinite set of real numbers (and a countably infinite set of rationals—in either case, a large set of numbers). Moreover, floating-point operations are not associative, unlike their counterparts over the reals or rationals. For example, unlike the integers or reals, floating-point numbers are not always associative under addition or multiplication. That is for floating-point numbers a, b, c, it is not always true that $(a + b) + c = a + (b + c)$. We shall consider these issues at length in Chapter 5. It is evident that **N** is embedded in **Z**, denoted $\mathbf{N} \subset \mathbf{Z}$. Similarly, $\mathbf{Z} \subset \mathbf{Q} \subset \mathbf{R}$. The operations defined over these sets are the familiar arithmetic ones (although their meaning may differ from set to set). We shall assume these operations are familiar.

Exercise 1.1. Recall from the previous chapter that for any $n > 0$, $P(n)$ was the sum

$$P(n) = \sum_{i=0}^{n} \frac{(-1)^i}{2i + 1}.$$

In the previous paragraph, we assumed each $P(n)$ is a rational number. Why is this true?

It is important to be able to construct pairs, triples, or more generally, n-tuples of the above sets. For example, we denote ordered pairs of reals, denoted $\mathbf{R} \times \mathbf{R}$ or $\mathbf{R}^2$, by the notation (x,y), where x, y are both in **R**. Occasionally it will be useful to identify the individual components of a point $P \in \mathbf{R}^2$ (or $\mathbf{Z}^2$ or $\mathbf{N}^2$) as x_P and y_P. We may also sometimes abuse notation in expressing the addition of offset a to x_P and b to y_P by $P + (a,b)$, where the addition is taken to be componentwise.

A closed interval $[a,b]$ over **R** (or **Z** or **N**) is the set $\{c \in \mathbf{R} : a \leq c \leq b\}$. A semi-closed (or semi-open) interval $(a,b]$ is $[a,b] - \{a\}$; similarly $[a,b)$ does not include b. The open interval (a,b) contains neither a nor b, but we shall use this notation with care because it conflicts with the use of brackets to denote ordered pairs. Products of intervals are used to create multi-dimensional boxes as in, for example, a rectangle $[a,b] \times [c,d]$. A unit cube is $[0,1] \times [0,1] \times [0,1]$, or $[0,1]^3$.

1.2. Vectors

Many quantities can be measured or reported using just a single number: temperature, density, time, height, distance, and so on, can be represented as a single real value $x \in \mathbf{R}$. These are called *scalar* quantities, or just *scalars*. There are many other phenomena that cannot be described by a single scalar: a position in space, for example, which as we saw is denoted by an ordered n-tuple such as $(x,y,z) \in \mathbf{R}^3$. Each of the x, y, and z are themselves scalars, so we see that composite quantities can be built from simple ones. Some phenomena even need an infinite set of numbers to represent them: a colour has a spectral representation indicating the amount present of each wavelength (or frequency) of light. The set of wavelengths in fact forms a continuum. Often, however, a specific colour can be approximated using a small tuple of between three and seven numbers, so that it can be thought of as a point in a *colour space*, just as a spatial position can be represented as a point in a Euclidean space. As we shall see later, a Fourier or Taylor series can also be thought of as the representation of a function involving an infinite set of numbers. We shall see that a polynomial can be thought of as a point of finite dimension in a "polynomial space."

Some phenomena that we may wish to measure are innately directional: wind, water flow, forces, and velocities are examples. To describe such phenomena, we need to represent the *direction* of travel and the extent or *magnitude* of travel in that direction. A *vector* is used to represent these two quantities. Many vectors can describe the same phenomenon. For example, we may say that we are travelling north-east at a speed of 90 km/hour. This, in fact, is a specific description of our *velocity*. The speed at which we are travelling is the magnitude or *length* of our velocity vector. We can change the units of speed arbitrarily. We may also describe direction differently. If, for example, we align a co-ordinate system with the positive x axis meaning east, the positive y axis meaning north, and the positive z axis meaning elevation, then travelling "north-east" at a speed of (about) 90 km/hour could be described as the vector $\mathbf{v} = (64, 64, \pm 0.001)$, where the third component corresponds to large potholes in the road! Normally we shall use bold lower-case letters to denote vectors and italic lower case letters to denote scalars. The magnitude of a vector $\mathbf{v} = (x,y,z)$ is defined as

$$\|\mathbf{v}\| = \sqrt{x^2 + y^2 + z^2} \ .$$

Points and vectors have the same representation as a tuple of scalars in $\mathbf{R}^n$. This can be confusing, and we may sometimes add to the confusion ourselves by sometimes treating points as vectors. The important thing to note is that vectors are intrinsically directional; they are not rooted at some point in space.

Exercise 1.2. Look at the nearest wall, and let us assume its surface is planar, not curved. How many vectors are perpendicular to it with a magnitude of one meter? Hint: more than zero and fewer than you may at first think. Explain.

There are some basic mathematical operations involving vectors. If we let $\mathbf{u} = (u_1, u_2, u_3)$, $\mathbf{v} = (v_1, v_2, v_3) \in \mathbf{R}^3$ be vectors and if $a, b \in \mathbf{R}$ are scalars, then

$$a\mathbf{u} = (au_1, au_2, au_3),$$

$$\mathbf{u} + \mathbf{v} = (u_1 + v_1, u_2 + v_2, u_3 + v_3).$$

Exercise 1.3. The two above statements can be collapsed into one by defining the effect of $a\mathbf{u} + b\mathbf{v}$. What should this result be?

The multiplication of two vectors is defined in two ways via an *inner product* and a *cross product*. The first results in a scalar while the second results in a vector. We will define inner product in Appendix C, but we will not require cross products in this book.

The *zero vector*, which is $(0,0,0)$ in $\mathbf{R}^3$, is degenerate in the sense that it has no magnitude and no direction. Two vectors are *parallel* if they have the same direction, but they can have different magnitude. In other words, $\mathbf{u}$ and $\mathbf{v}$ are parallel if there are non-zero scalars a, b such that

$$\mathbf{u} = a(x,y,z), \quad \text{and}$$

$$\mathbf{v} = b(x,y,z),$$

for some non-zero vector (x,y,z). A *unit vector* is one that has magnitude one. Any non-zero vector $\mathbf{u}$ can be *normalised* into a unit vector $\hat{\mathbf{u}}$ by dividing the components of $\mathbf{u}$ by its magnitude:

$$\hat{\mathbf{u}} = \frac{1}{\|\mathbf{u}\|}\mathbf{u}.$$

Appendix C briefly discusses matrix representations related to vectors.

1.3. Functions

A *function f* maps elements of one set A to elements of some other set B. A typical shorthand for this is $f: A \to B$, and of course our choice of the labels A, B, and f is irrelevant. A is called the *domain* of f, and B is a *target* of f. We can extend f to operate over a set of points $A' \subseteq A$ by defining $f(A')$ to be the set $\{f(x): x \in A'\}$. The *range* of f is, naturally enough, the set of values of B that $f(x)$ actually varies over, for $x \in A$. Formally, the range of f is the set $f(A)$. Notice, therefore, that the range of f is a subset of a target: a target can have values in it to which f never refers, while by definition the range of f can contain only the values to which f refers.

Several properties of functions must also be defined. A function $f: A \rightarrow B$ is *one-one* or *injective* if f maps distinct elements of A to distinct elements of B. That is, if $a_1, a_2 \in A$ and $a_1 \neq a_2$, then $f(a_1) \neq f(a_2)$.

A function $f: A \rightarrow B$ is *onto* or *surjective* if every point $b \in B$ has a representative $a \in A$ such that $f(a) = b$. In this case, then, the range of f is the same as its target. If f is both injective and surjective, then f is called *bijective*. We also expect functions $f(x)$ to denote a single value y, and not a set $\{y_0, y_1, \cdots, y_n\}$. Because a circle can have two values in y for each x (and vice versa), it is technically a *curve*, not a function. We shall return to this point.

Exercise 1.4. The function $f: \mathbf{R} \rightarrow \mathbf{R}$ given by $f(x) = 2$ is neither one-one nor onto. Why?

Example 1.1. Consider the function $f: \mathbf{R} \rightarrow \mathbf{R}$ given by $f(x) = x^2$. A graph of this function is given in Figure 1.1. If we draw a horizontal line $y = c$ for some $c > 0$, then we see that it makes two intersections with f. Thus f cannot be one-one, since two distinct x values map to the same y. Furthermore, f is not onto, since $f(x)$ is never less than zero. If, however, we restrict the domain of f to be the positive reals, then f is one-one. Moreover, if the target is restricted to the positive reals (including zero), then f is surjective. The moral of this story is that the choice of domain and target can greatly affect the classification of a function.

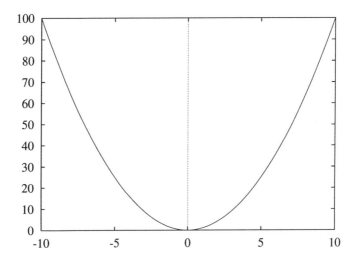

Figure 1.1.   The function $f(x) = x^2$ on $[-10, 10]$.

If $f: A \rightarrow B$ and $g: B \rightarrow C$ are both functions, then the *composition* of g with f is another function $h: A \rightarrow C$ given by $h(x) = g(f(x))$. This is sometimes written $h = g \circ f$. Computationally, this means that to determine the value of $h(x)$, we first evaluate the function f at x to determine $y = f(x)$, and then we evaluate g at y to determine $h(x) = g(y)$.

If $f: A \rightarrow B$ is a function, then its *functional inverse*, $f^{-1}: B \rightarrow A$, if it exists, is such that $f(f^{-1}(x)) = f^{-1}(f(x)) = x$. If a function is bijective, then it has an inverse (and vice versa).

Exercise 1.5. Taking $f: A \rightarrow B$ and $g: B \rightarrow C$ as above, under what conditions is the function $h = f \circ g$ meaningful? If $f: \mathbf{R} \rightarrow \mathbf{R}$ is defined as $f(x) = \log_2(x)$ and $g: \mathbf{R} \rightarrow \mathbf{R}$ is $g(x) = x^2$, is $f \circ g$ meaningful? Is $g \circ f$ meaningful? (Careful! What is the true range of f?)

Exercise 1.6. Consider the function $f: A \rightarrow B$ given by $f(x) = x^2$, $A, B \subseteq \mathbf{R}$. What is the inverse of f, and what are the constraints on A, B for this inverse to be valid? In general, what is the inverse of each function $f: A \rightarrow B$ when $f(x) = x^k$, for any natural number k?

A function $f: A \rightarrow B$ is *continuous* at a point $\hat{x} \in A$ if for any sequence $x_0, x_1, \cdots$ that converges to $\hat{x}$, the corresponding sequence $f(x_0), f(x_1), \cdots$ converges to $f(\hat{x})$. This is one of many ways of defining a continuous function and is the most compact for our purposes. The common intuition for defining a continuous one-dimensional function is that it can be drawn without lifting your pen. We can also say that a function is continuous at a point $\hat{x}$ if for any point x in a small region close to $\hat{x}$, the value $f(x)$ is close to the value of $f(\hat{x})$. The function $y = x^2$ is continuous for any $x \in \mathbf{R}$, as is the function $y = |x|$, where $|x|$ denotes the magnitude or absolute value of x. However, the function $y = 1/x$ is *not* continuous at $x = 0$. A function that is not continuous is *discontinuous*. A function is continuous on a set A if it is continuous at every point $x \in A$. Thus $y = x^2$ is continuous on $[-1,1]$, but $y = 1/x$ is not.

Exercise 1.7. Graph the functions $y = |x|$ and $y = \max(x, 1)$, for x in $[-5,5]$.

Exercise 1.8. Is the function $y = 1$ continuous on $\mathbf{R}$?

Exercise 1.9. The function $f(x) = (\sin x)/x$ is well defined for every value of x in $\mathbf{R}$ but one. Use Maple to investigate the behaviour of $f(x)$ near this value and deduce whether or not f is continuous for all x in $\mathbf{R}$.

Exercise 1.10. Do the same for $f(x) = \sin(1/x)$? Is your conclusion the same?

The notion of a *curve* is very clear to us intuitively: we imagine a curve as a connected sequence of points in space. A circle or the trajectory of an object in flight are both curves. The exact definition of a curve is in fact surprisingly delicate, but for our purposes it will suffice to think of a curve as a continuous mapping (or deformation) of an interval on the real line. That is, we can imagine getting any curve we like by pulling, twisting and moving a very flexible and stretchy line segment. We similarly define a *surface* as being a continuous mapping of a rectangle lying in $\mathbf{R}^2$. In this case, think of the rectangle as being a rubber sheet that can be easily manipulated to form an arbitrary curved surface. In this book, we will represent all curves and surfaces in terms of simple functions, and we will sometimes blur the distinction between functions and curves (as perhaps the title of this chapter suggests).

1.4. Representation of Functions, Curves, and Surfaces

1.4.1. Explicit and Implicit Representations

There are many equally precise ways of writing down the same function. This enriches our mathematical vocabulary, since we can use the notation that best suits our needs for the task at hand. However, it is sometimes not obvious when two different definitions actually denote the same object.

The most common way of defining a function is to do so *explicitly*, meaning that a variable that represents the value of the function is placed on one side of an equation, and an expression not containing that variable is placed on the other side of the equation. To define a line, for example, we could write

$$y = f(x),$$
$$= mx + b,$$

with m being the slope of the line, and b being the y-intercept. For an arbitrary parabola, we could write

$$y = ax^2 + bx + c.$$

More generally, f could be a function of many variables, and the values that f takes on could be tuples. An explicit formulation is particularly useful when one is always solving for the dependent variable y.

Exercise 1.11. A ball of mass m is dropped from a height of $y = y_0$ at time $t = 0$. Write down an explicit function that gives the vertical height of the ball at time $t > 0$.

It is easier to define curves and surfaces *implicitly*. In this case, no variable is isolated as in the explicit case. We usually write an implicit equation as

$$F(x,y) = 0.$$

For example, the equation of a circle of radius $r > 0$ centred at the origin is

$$F(x,y) = x^2 + y^2 - r^2.$$

The surface of a sphere of radius r centred at the origin has the implicit form:

$$F(x,y,z) = x^2 + y^2 + z^2 - r^2. \tag{1}$$

The equation of an arbitrary plane embedded in $\mathbf{R}^3$ is

$$F(x,y,z) = Ax + By + Cz + D,$$

where $A, B, C, D \in \mathbf{R}$ are constants (for a given plane). As can be seen, implicit functions are particularly useful in characterising simple geometric objects. It is usually (but not always) straightforward to convert between explicit and implicit functional representations.

Exercise 1.12. Write down a formula for a general line in implicit form (not easy).

Exercise 1.13. Convert the implicit equations for the sphere and the plane into explicit functions of the form $z = f(x,y)$.

1.4.2. Parametric Representations

Suppose we wish to describe a complex trajectory of a flying object in terms of a curve. We need functions that tell us, at each instant of time t, the position of the object. We can do this implicitly or explicitly, but since the object can perform death-defying loops and spins, such functions can become extremely complicated. Another possibility would be to simplify the problem by trying to write down the progress made by the object in each dimension x, y, and z with respect to time t in the hope that the behaviour of the object in each independent dimension is simpler than the overall behaviour. To restate things, rather than writing $z = f(x,y)$, or $F(x,y,z) = 0$, we can instead try to find three separate functions

$$x = f_x(t),$$

$$y = f_y(t),$$

$$z = f_z(t).$$

The resulting curve

$$\mathbf{F}(t) = \left(f_x(t), f_y(t), f_z(t) \right)$$

describes, as t varies, a path of points in $\mathbf{R}^3$ that the object follows. We have just defined a curve in terms of three *parametric* functions in a single *parameter t*. We could also say that we have *parameterised* the trajectory of the object. We write a parametric curve $\mathbf{F}$ using a bold font to make clear the fact that $\mathbf{F}$ is actually a *tuple* of functions. It is legitimate to view $\mathbf{F}$ as a (position) *vector*.

The use of parametric curves is a powerful tool, and we shall explore some examples of their power in this section. The first thing to notice is that we can immediately depict an approximation to the above trajectory. If our simulation of the trajectory is to last T seconds starting at $t = 0$, and we are to plot points every Δt seconds, then the pseudocode in Figure 1.2 gives a visual approximation to the trajectory. Notice, as well, that it is easy for us to change the behaviour of the function in one dimension without affecting the others. For example, in our rendering we could scale the y dimension to be ten times that of the others with the following replacement to the starred line above:

$$\texttt{to} := \left(f_x(t), 10f_y(t), f_z(t) \right).$$

Figure 1.3 illustrates that uniform steps of Δt in the domain for parameter t does not necessarily produce equal-length arcs along $\mathbf{F}(t)$. Also notice our earlier suggestion that a curve can be thought of as the deformation of a line segment.

```
const T: real := NumberSeconds
const Δt: real := deltaT
var t:   real := 0
var from, to: Point
```

from := $(f_x(0), f_y(0), f_z(0))$

loop

    $t := t + \Delta t$

    exit when $t >$ **T**

    to := $(f_x(t), f_y(t), f_z(t))$ *

    PutLine(from,to)

    from := **to**

end

Figure 1.2. An approximation to the trajectory in terms of line segments.

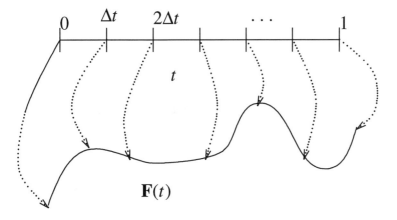

Figure 1.3. Mapping uniformly-spaced parameter values to curve values.

When we look at parametric functions with more than one parameter, we shall see that it is just as easy to render higher-dimensional objects. Parameterisation allows us to decompose a surface into approximate, geometrically simpler objects such as triangles and quadrilaterals. This has applications not just to computer graphics, but to any mathematical application that requires computations to be performed over an object's surface. Sample applications in which this sort of operation is performed include heat transfer, stress analysis, and indeed any application involving the finite-element method.

Parametric definitions offer excellent control over an object, particularly if the domain of the parameter is well-chosen. For example, there are many ways to represent a line or line segment. Given a line segment with endpoints (P_0, P_1), a compact parametric representation is

$$\mathbf{L}(t) = (1-t)P_0 + tP_1.$$

This is an abbreviation for the equations:

$$x(t) = (1-t)x_0 + tx_1,$$
$$y(t) = (1-t)y_0 + ty_1.$$

For a line segment in $\mathbf{R}^3$, $\mathbf{L}$ would describe an additional equation:

$$z(t) = (1-t)z_0 + tz_1.$$

In all cases, the parameter t varies over [0,1]. Another version of a line segment that is more economical in the use of the parameter t is

$$\mathbf{L}(t) = P_0 + t(P_1 - P_0).$$

Exercise 1.14. Verify that for $t = 0$, $\mathbf{L}(0) = P_0$, and that for $t = 1$, $\mathbf{L}(1) = P_1$.

Exercise 1.15. Show that $\mathbf{L}(\frac{1}{2})$ gives the midpoint of the line segment.

Exercise 1.16. Show that varying t by n equally sized steps of size Δt gives equally spaced points $\mathbf{L}(i\Delta t)$, $i = 0, 1, 2, \cdots, n$. Assume n is such that $n\Delta t = 1$.

Many different parameterisations exist to describe the same line segment. For example, if we replace t by $(u + 1)/3$ and substitute into $\mathbf{L}(t)$ above, we get the following equation after simplifying:

$$\mathbf{L}(u) = u\,\frac{P_1 - P_0}{3} + \frac{P_1 + 2P_0}{3}.$$

This gives exactly the same line segment as above, except that we must determine the domain for the new parameter u. Since

$$t = \frac{u+1}{3},$$

$u = 3t - 1$. Thus when $t = 0$, $u = -1$, and when $t = 1$, $u = 2$, giving a domain of $[-1,2]$. When a parametric function is converted to another parametric function in this manner, the process is called *reparameterisation*.

Example 1.2. Suppose we let t be an arbitrary function $f(u)$. Consider its effect on $\mathbf{L}(t)$:

$$\mathbf{L}(t) = \mathbf{L}(f(u)) = P_0 + f(u)(P_1 - P_0).$$

In some cases, such as when $f(u)$ is a polynomial, we can rewrite this equation directly in terms of u. We must ensure that the required domain, in this case [0,1], of t is preserved by finding the u for which $f(u) = 0$ and $f(u) = 1$. In general, we may require the techniques described in Chapter 6 for this. The following Maple procedure reparameterises one component of $\mathbf{L}(t)$, replacing t by $f(u)$, such that $u \in \mathbf{I} = [a,b]$.

```
reparamline := proc(x0,x1,f,I)
   local t, L, L1, u;

   L  := x0 + t*(x1-x0);              # original segment
   L1 := simplify(subs(t=f, L));      # reparam of segment

   u  := indets(f,name)[1];           # the variable of f
   print(plot( [f,L1, u=I[1]..I[2]])); # plot L1 as fn of f
   print(plot( L1, u=I[1]..I[2]));     # plot L1 as fn of u
end:
```

L is plotted both as a function of u and as a function of $f(u)$. For *any* non-zero $f(u)$, the latter plot should be a line segment. How should the former plot look for any f? What does it represent? As an example, **reparamline(1,4, 3*u, [0,1/3])** should give the same line segment as **reparamline(1,4, u, [0,1])**.

Exercise 1.17. If $t = u^2$, give the two intervals in u over which **L**(t) traces the desired line segment. Does each cause points on the line segment to be evaluated in the same order?

Exercise 1.18. Compute the relevant new domain in u if $t = e^u$, $t = e^{-u}$, $t = 1/u$. Verify this using **reparamline**.

Exercise 1.19. Use the Maple function **solve** to change **reparamline** so that the argument **I=[a,b]** now represents the domain of t, so that you now must solve for the value(s) of u for which $f(u) = a$ and b.

A parameter can represent anything. For us, it often represents time. Also observe from the previous example that we can reparameterise the line segment and change the effective *speed* at which the curve is drawn. In particular, if we step through a parameter at a constant spacing, say Δt, then different parameterisations all give the same parametric function **P**, but they change the spacing between evaluations of **P**$(i\Delta t)$.

Let us consider two parametric definitions of a circle with centre (x_c, y_c) and radius r (see Figure 1.4). The first is

$$x(\theta) = x_c + r\cos\theta,$$

$$y(\theta) = y_c + r\sin\theta,$$

where $0 \leq \theta < 2\pi$. We should stress two simple, important points. First, because the sine and cosine functions are periodic, a wide variety of domains would give the same circle, such as $[-\pi, +\pi]$, and $[-2\pi, 0]$. Furthermore, the same circle can be traced out several times if we make our domain larger. For example, if θ is allowed vary over $[0, 8\pi]$, then the same circle would be traced out four times. Second, as suggested earlier, while $x(\theta)$ and $y(\theta)$ are true functions, the circle

$$\mathbf{C}(\theta) = (x(\theta), y(\theta))$$

is not a function in terms of rectangular or Cartesian co-ordinates (x, y): each x can have zero, one or two distinct values for $f(x)$. We shall let the term *curve* denote an object such as a circle whose graph may not be a function but whose

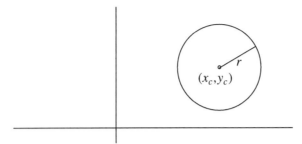

Figure 1.4.   Circle with centre (x_c, y_c) and radius r.

co-ordinates can be written down in terms of parametric functions. Curves in $\mathbf{R}^n$ are also called *space curves*.

Exercise 1.20. Suggest a definition for *continuity* when dealing with parametric curves such as $\mathbf{C}(\theta)$ instead of functions.

Example 1.3. The following Maple code plots a circle of radius 1, together with its x and y components of the parametric function. These functions are plotted along their co-ordinate axes with respect to θ.

```
x := cos(theta):
y := sin(theta):
plot({ [theta,x,theta=0..2*Pi], [y,theta,theta=0..2*Pi],
    [x,y,theta=0..2*Pi]});
```

The result is depicted in Figure 1.5. What point on the circle corresponds to $\theta = 0$?

Exercise 1.21. Consider the parametric space curve $\mathbf{C}(t) = \big(x(t), y(t), z(t)\big)$ given by

$$x(t) = \cos t,$$
$$y(t) = \sin t,$$
$$z(t) = t,$$

for $t \in [0,6\pi]$. If we "look" at $\mathbf{C}(t)$ from along the z axis, then we get a unit circle as above. From a general viewpoint, however, we get a spiral as depicted in Figure 1.6. Write a Maple program to plot this spiral.

Another parametric representation of a circle of radius 1 with centre $(0,0)$ is

$$x(t) = \frac{1-t^2}{1+t^2},$$

$$y(t) = \frac{2t}{1+t^2}.$$

The functions $x(t)$ and $y(t)$ are examples of *rational polynomials*, namely fractions with polynomial numerators and denominators. They express conics (like circles and ellipses) that cannot be expressed using non-rational polynomials.

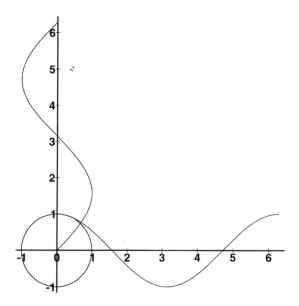

Figure 1.5. A circle together with its parametric functions plotted along their respective co-ordinate axes.

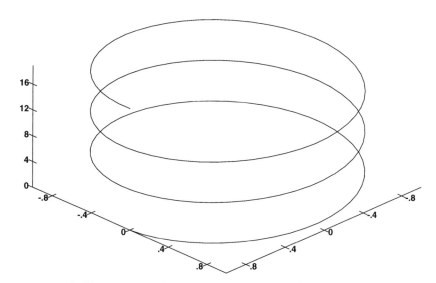

Figure 1.6. A spiral space curve.

Exercise 1.22. Plot the rational representation of the circle using Maple for $t \in [0,1]$. What part of the circle is drawn? What must t vary over to parameterise a whole circle? This demonstrates that a parameter need not be defined over a finite interval to define a "finite" object.

Exercise 1.23. Starting from the trigonometric parameterisation, derive an explicit function, containing no trigonometric functions, describing a circle of radius r and centred at (x_c, y_c). Hint: derive expressions for $\sin u$ and $\cos u$, square both equations, and add.

Exercise 1.24. Show that by making a very simple change to the parametric representation of a circle, we can get all ellipses. Use Maple to plot several different ellipses.

Example 1.4. A *trochoid* has parametric representation:

$$x(u) = au - r \sin u,$$
$$y(u) = r \cos u.$$

The values a and r are real constants, and define a different trochoid for each a and r. The curve resulting from $a = 0.75$ and $r = 3$ is depicted in Figure 1.7. Unlike the definition of a circle, notice that as the domain over which the trochoid is plotted increases, the range of the curve in x will correspondingly grow, while the range in y will remain the same.

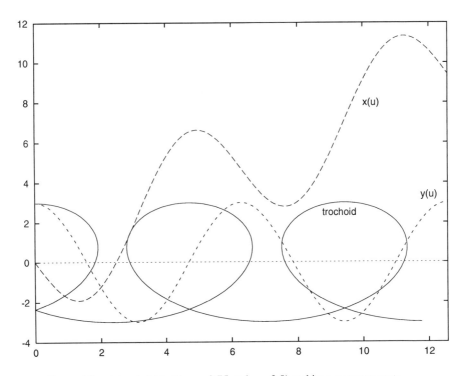

Figure 1.7.   A trochoid (with $a = 0.75$ and $r = 3.0$) and its x, y components.

Exercise 1.25. Use Maple to plot the following two sets of parametric functions:

$$x(\theta) = \frac{\sin \theta}{\theta},$$

$$y(\theta) = \sin \theta,$$

and the other with x as above, but with $y = \cos \theta$. Plot these functions for fairly large intervals in θ that are symmetric about zero, such as $[-12\pi, 12\pi]$. Explain the results.

To determine the dimension of the object described by a parametric representation, we just count the number of independent parameters. In all of our previous examples, one parameter was used, which means were were describing curves. The number of components in the parametric definition gives the dimension of the space in which the object is embedded. Curves embedded in the *xy*-plane have two parametric equations. Curves that can twist through three-dimensional (3-D) space would have a third component. Equations of the form

$$x = f_x(u,v),$$

$$y = f_y(u,v),$$

$$z = f_z(u,v),$$

describe a two-dimensional *surface*, because the object is a function of two parameters u and v (assuming they are both used). In this case, the surface is embedded in a 3-D space. A parametric definition of the surface of a sphere is

$$x(\theta, \phi) = x_c + r \sin \phi \sin \theta,$$

$$y(\theta, \phi) = y_c + r \sin \phi \cos \theta, \tag{2}$$

$$z(\phi) = z_c + r \cos \phi,$$

with angles θ and ϕ such that $0 \le \theta < 2\pi$ and $-\pi \le \phi < \pi$. Consider Figure 1.8. The lines of latitude and longitude correspond to evaluations of the parametric functions x, y, and z in which one parameter varies and the other is held constant. Such curves are sometimes called *isoparametric* curves. This is a useful way of thinking about the parameterisation. Let us put the centre of the sphere at the origin. If we let ϕ remain constant and let θ vary, then z is constant, and the $r \sin \phi$ term in x and y is constant. The result is a circle of radius $r \sin \phi$ in the $z = r \cos \phi$ plane slicing through the sphere. For each value of ϕ, we get a circle of a different radius drawn in a different plane of z. If we instead think of ϕ as varying and of θ as being constant, then we get similar circles in the x and y planes.

Exercise 1.26. Give similar parametric equations for the cylinder in Figure 1.9 and plot them using Maple. Show how a simple change of the equations results in a cone with apex at the origin and growing with increasing z. Then derive implicit formulae for both the cylinder and the cone. Can you easily make the sides of the cone curved as z increases? In other words, show how to go from the cone in Figure 1.10 at left to the "flower" at right.

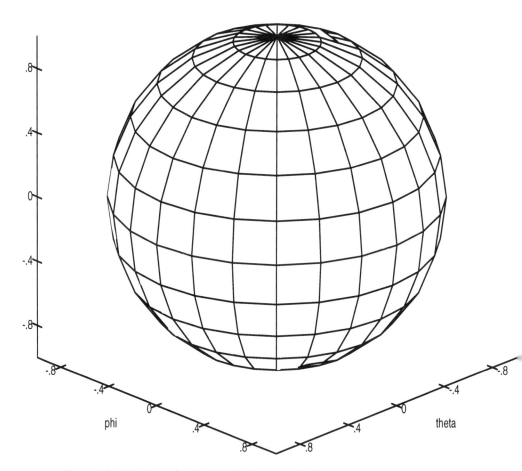

Figure 1.8.   A sphere of radius 1 defined parametrically.

Parametric representations can be extremely powerful for *modelling* geometric objects. In Figure 1.11, we see a geometric model of a teapot. The details of its construction are unimportant for now, but notice that this object is considerably more complicated than the algebraic objects we have seen so far: the surfaces vary greatly in smoothness, the *topology* of the shape changes, and it is constructed from *pieces* both visible and invisible. The visible pieces are the teapot lid, and the teapot body; the spout and handle can be thought of as independent pieces. In fact, even the smooth teapot body consists of many curved parametric surface *patches* that are put together in an invisible *piecewise* fashion. We shall see many examples of these ideas shortly.

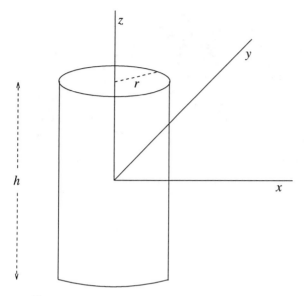

Figure 1.9.   Cylinder with height h and radius r.

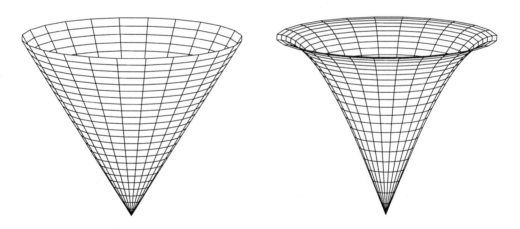

Figure 1.10.   A cone and a flower.

We have seen that if we know the domain of a parametric representation, then we can easily compute, for any point in the domain, the corresponding point on the curve or surface. However, it is in general not at all easy to go the other way: determining the value of the parameter from a point on a surface is often not easy, even if our representation is bijective.

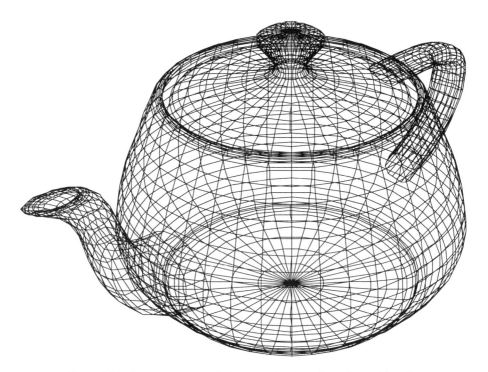

Figure 1.11. A teapot composed of parametric piecewise polynomial surfaces.

Exercise 1.27. Explain this last statement. Given a point P on the surface of a sphere defined trigonometrically as in Eq. 2, show how to solve for θ and ϕ. Is the problem any easier if we solve using an implicit representation as in Eq. 1?

1.4.3. Polynomial Representations

The examples above showed that a wide variety of functions can be used in parametric representations. We also saw that given one parametric representation, we can generate many others by reparameterisation: if our representation uses a parameter t, we simply replace all occurrences of t in our representation by a well-behaved function $f(u)$. While this can be a useful tool, it is generally best to use the simplest and least costly parameterisation of an object. In computer science, ''cost'' is measured in terms of storage requirements and in terms of the *complexity* of computation. *Polynomial* functions satisfy these criteria well, and are sufficiently flexible to express many of the shapes we want and to approximate most functions very well.

A polynomial $p(t)$ with real coefficients $a_0, a_1, \cdots, a_n$ as a function of parameter t is of the form

$$p(t) = a_0 + a_1 t^1 + a_2 t^2 + \cdots + a_n t^n. \tag{3}$$

The polynomial is of *degree n* if the highest power in t with a non-zero coefficient a_n is n. The *order* of $p(t)$ is the number of possible terms in $p(t)$, and hence the order of an nth-degree polynomial is $n+1$. To *evaluate* a polynomial, that is to compute $p(t)$ for any desired t, only additions, subtractions, and multiplications are needed. Divisions are needed only for *rational polynomials*, that is, functions of the form $p(t)/q(t)$, for polynomials p, q. Earlier, we saw an example of rational polynomials that described a portion of a circle. Rational polynomials have recently become very popular in computer graphics.

Now we can address the issue of the "cost" of a parametric representation. The order of a polynomial directly determines its storage cost: if the polynomial is of order $n+1$, then we know that at most $n+1$ numbers are needed to identify uniquely the polynomial. Four numbers, for example, will uniquely determine a cubic polynomial. Let us consider the cost of evaluating various kinds of functions. Table 1.1 gives the time required to make 1 000 000 evaluations of various functions. This test was run on a Sun SPARCstation 2 using the built-in floating-point accelerator. The computations were performed in double precision.[1] Programs were written in the C programming language and were compiled using the standard C compiler without optimisation and with the cost of loop control removed. As we shall see later, there are several methods of reducing the cost of evaluating polynomials beyond the direct technique. Two such techniques are Horner's rule and forward differencing. In C, a polynomial of degree n would require $n(n+1)/2$ multiplications and n additions if we were to evaluate each term independently. This is clearly not efficient.

The data in our table clearly support the contention that polynomials can be efficiently computed. A warning is in order, however. Most common mathematical functions are implemented in hardware and increasingly, chips that implement mathematical functions run at speeds that exceed the speed of the computer's central processing unit. In the future we are very likely to see the gap close between the evaluation of polynomials and other functions such as those found in the table.[2] As it stands, several low-order polynomials can be evaluated for the cost of a single transcendental function. We shall take advantage of this efficiency, as we often put polynomials together to model interesting objects.

It is worth remarking in passing that Eq. 3 is a *specification* of a function p that associates, for any given value t in the relevant domain, a value $p(t)$ to t. We have already suggested that there are at least three possible functionally equivalent *implementations* or *realisations* of this specification: direct evaluation,

---

1. Floating point representations will be discussed in Chapter 5.

2. This presumes that hardware support for polynomial evaluation would not be available, which is probably not true. As we shall see later, a hardware matrix multiplication unit can be used to evaluate several points on a polynomial curve very quickly and in parallel.

Function	Computation Cost (seconds)
cubic (direct evaluation)	1.25
cubic (Horner's rule)	0.82
cubic (forward differencing)	0.38
quartic (direct evaluation)	1.93
quartic (Horner's rule)	1.07
quartic (forward differencing)	0.55
e^x	4.77
$\sin x$	5.95
$\sqrt{x}$	7.61

Table 1.1.   The cost of computing various mathematical functions. Each function
was computed 1 000 000 times.

Horner's rule, and forward differencing. However, of all those implementations
that successfully achieve the specification, we are most interested in those that do
so the quickest. Furthermore, implementations on real computers are usually only
equivalent if exact arithmetic is presumed, so that the issue of *accuracy* and
stability must also be taken into account when deciding which implementation is
the best. On balance, there is usually no single best approach.

Returning to our discussion of polynomials, we can make things simpler for
ourselves with respect to the domain of the parameter by noting that if a parameter varies over any finite interval $[a,b]$, we can reparameterise the polynomial;
this action gives another polynomial of the same degree, such that the same curve
or surface is described by a parameter with domain $[0,1]$. Suppose a polynomial
on parameter t is of interest over the interval $t \in [a,b]$, $a < b$. If we replace t by

$$u = \frac{t-a}{b-a},$$

then observe that u varies over $[0,1]$ as t varies from a to b. Now, from this equation we can rewrite t as $ub - ua + a$, and subsequently replace each occurrence of
t in $p(t)$ by this expression in u. Upon simplifying, we obtain an equivalent polynomial $\hat{p}(u)$ with $u \in [0,1]$. This is an excellent example of the power of
reparameterisation.

Exercise 1.28. Verify that $\hat{p}(u)$ is of the same degree as $p(t)$.

Example 1.5. Suppose $p(x) = x^2 - 1$ is of interest to us over the interval $[-2,2]$. To
reparameterise $p(x)$ to a new polynomial $\hat{p}(t)$ such that $t \in [0,1]$, note that $t = (x+2)/4$,
and so $x = 4t - 2$. Therefore,

$$\hat{p}(t) = p(4t-2) = (4t-2)^2 - 1 = 16t^2 - 16t + 3.$$

Exercise 1.29. Using pen and paper or using Maple, plot the polynomial

$$p(x) = 3x^3 - 2x^2 + 4x - 1$$

over the interval $[-2,2]$. Then reparameterise $p(x)$ to $\hat{p}(t)$ where t ranges over $[0, 1]$. Write down the polynomial $\hat{p}(t)$ in simplest form.

What degree of polynomial is the best? The answer of course depends on the application, and we shall discuss the issue of polynomial degree at length later. Degree zero polynomials are constant functions, which will probably not be of much use. Degree one polynomials are lines or line segments, which provide us with the ability to interpolate between two values in $\mathbf{R}^n$. Polynomials of degree two are quadratics in their parameter. Consequently, space curves of degree two are parabolic arcs, and are also *planar*, meaning that the curve is embedded in a plane within $\mathbf{R}^n$. For example, consider the quadratic space curve $\mathbf{C}(t) = (x(t), y(t), z(t))$ given by

$$x(t) = 5t^2 - 2t + 1,$$
$$y(t) = -5t^2,$$
$$z(t) = 2t^2 + 4t.$$

Figure 1.12 depicts this space curve for $t \in [-1.5, 1.5]$.

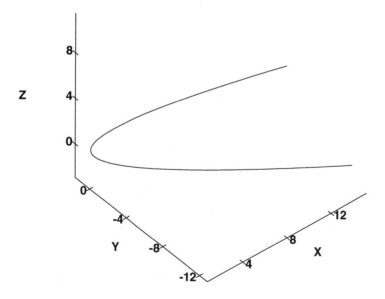

Figure 1.12. A parabolic space curve.

If we change $x(t)$ to

$$x(t) = 16t^3 - 5t^2 - 2t + 1,$$

leaving the quadratic functions $y(t)$ and $z(t)$ as before, we get a non-planar curve. Figure 1.13 depicts the curve for $t \in [-3, 2]$.

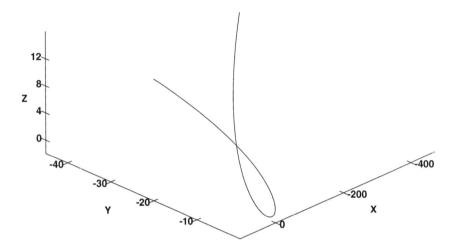

Figure 1.13. A space curve with one cubic and two quadratic components.

Finally, letting y and z also be the following cubic functions:

$$y(t) = 12t^3 - 8t^2,$$
$$z(t) = 8t^3 - 2t^2,$$

we get a space curve that is both non-planar and that has a point of inflection (i.e., a point at which the second derivative is zero). Figure 1.14 illustrates this curve for $t \in [-0.5, 0.5]$.

A particularly attractive feature of polynomial representations is that while individual parametric functions are fairly uninteresting, sequences of polynomials can be ''stitched'' together to form more complicated and interesting space curves (or surfaces). A function defined by sequences of polynomials is said to have a *piecewise polynomial* representation. This property is extremely important to many applications, and we shall deal extensively with piecewise representations in this book.

Although we shall consider this issue at length in the next chapter, it is important to note at this juncture that many distinct representations exist even for polynomials as simple as lines. As we have seen, one way to introduce arbitrarily

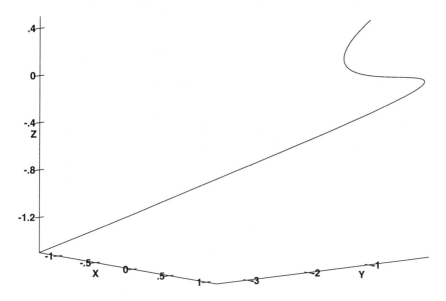

Figure 1.14. A cubic space curve.

many representations is through reparameterisation. However, even when the same parametric domain and "observables" such as endpoints are retained, we can devise different representations for the same class of polynomials. For example, the representation

$$\mathbf{L}(t) = (1-t)P_0 + tP_1$$

causes us to think about a line segment as a weighted sum of two endpoints. In contrast, the expression

$$\mathbf{L}(t) = P_0 + t(P_1 - P_0)$$

instead makes us think of the same line segment as part of a ray rooted at P_0, travelling in the direction given by $P_1 - P_0$. Of course it is easy to write one representation in terms of the other, but this should not disguise the fact that each representation allows us to think of the line segment in a different way. In both cases, we need two pieces of information to define the line segment. As we shall see, each possible representation of a class of polynomials of order $n+1$ forms an $(n+1)$-dimensional function space, in which precisely $n+1$ pieces of information are required to represent uniquely a specific polynomial of degree n. Changing representations will amount to a *change of basis*, which allows us to convert from one way of writing $n+1$ pieces of information to another. This will become particularly important when we discover that there are a great many good representations for higher-order polynomials.

Exercise 1.30. Modify the pseudocode in Figure 1.2 to render three parametric functions of two parameters. Use this pseudocode to write a procedure in Turing or C that takes as arguments:

- three parametric functions $f_x(s,t)$, $f_y(s,t)$, and $f_z(s,t)$ of two real-valued variables (s,t).

- four real values, s_{min}, s_{max}, t_{min}, t_{max} denoting the endpoints of the intervals over which to render the parametric functions.

- two subdivision values N_s and N_t denoting the number of equally spaced intervals in s and t respectively at which to evaluate the parametric functions. There will therefore be $(N_s + 1)(N_t + 1)$ evaluations of each of f_x, f_y, and f_z. However, unless care is taken, many more evaluations than this will be required. Optimise your algorithm so that it evaluates the three functions at each sample point *exactly* once.

After writing this routine, test it by defining several Turing or C functions and pass them to your routine. Try out your parametric functions for the sphere, cone, and cylinder.

1.4.4. Procedural Representations

So far we have considered the mathematical definition of objects as consisting of a set of functions that define the behaviour of an object in space. In some cases, there is an complementary view in which the object is *generated* to an arbitrary level of detail. We will not take this notion very far in this book, but it is important to point out this alternative. We do so by means of an example.

In computer graphics and computer vision, it is often of interest to model objects that are irregular in appearance. Examples of this include terrain (the ground, mountains, elevation data, etc.), clouds, fire, and so on. It is often very difficult to come up with a suitable set of functions that describe the behavior of such objects. While there in fact are functional techniques for synthesising such phenomena using the Fourier transform methods described later in this book, it is often easier to describe the behaviour in terms of a set of rules that are repeatedly applied.

In this discussion, we will restrict ourselves to *deterministic* forms, that is, techniques that do not employ random processes.[3] Many recursive and iterated forms such as spirals, leaves, ferns, shells, branching structures, and goat horns can be characterised in this way. For example, suppose we work only with line segments and we specify a rule to transform a line segment to a sequence of line segments. One such rule can be found in Figure 1.15, in which a line segment of length 1 is transformed into four connected line segments, each of length $1/3$.

---

3. In fact, the generation of "random" numbers on a computer is actually itself a deterministic process, but this does not mean that the real process being modelled is deterministic. Every computational model of a continuous phenomenon is in some way an approximation.

Figure 1.15. A replacement rule: replace a line segment by a triangular shape.

By applying the replacement rule to all line segments in the current shape, we get a new shape. The only trick is that the replacement must be done in same orientation as the line being replaced. Figure 1.16 depicts a sequence of replacements starting with the original line segment and progressing to the classic *von Koch* curve. If we were to start with a different initial shape, such as a triangle, then after five iterations, we would get the resulting *von Koch snowflake* in Figure 1.17.

By changing the rule so that the triangular substitution always points inward, we get perhaps the most interesting shape depicted in Figure 1.18. These beautiful shapes, sometimes called *deterministic fractals*, are generated using simple replacement rules. There are endless variations on the rules and shapes that can be generated. How can we describe these rules? One way is to provide a *grammar*, which describes the replacement mechanism. Another way is to write a program to do the replacement. The Maple code in Figure 1.19 handles a general class of replacement rules, and includes the von Koch curves as a special case. It is fairly sophisticated, so the reader may first wish to get more familiar with Maple by consulting the appendix at the end of this chapter.

The procedure **Replace** replaces the current shape according to the replacement rule. **RepVonKoch** describes the replacement rule for the von Koch curves. Observe that it breaks a line segment up into four pieces as per the replacement rule described earlier. The rotation accomplishes the replacement in the appropriate orientation. To apply the replacement rule *n* times, we write:

```
g := Line:                   # or another initial value
for i from 1 to n do
    g := Replace(g,RepVonKoch):
od:
plot(g,axes=NONE,scaling=UNCONSTRAINED);
```

The study of models created by replacement, iteration, and possibly randomness is an immensely rich and rewarding subject area. We have considered only one example of a deterministic replacement process. There are many ways to extend this simple paradigm:

- The shapes can be generalised to higher dimensions to create surfaces and volumes.

Figure 1.16. A sequence of replacements, converging to the von Koch curve.

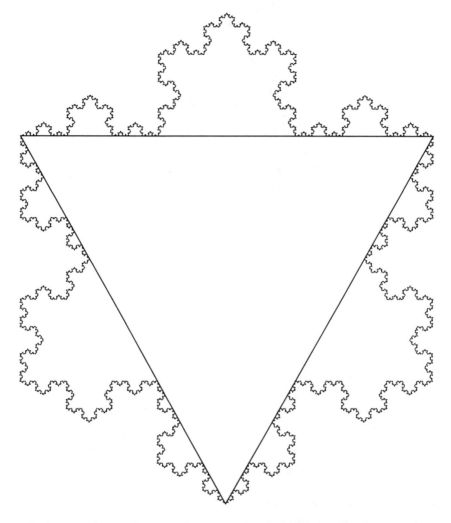

Figure 1.17. The von Koch snowflake embedding the initial triangular shape.

- The replacement patterns can be randomly perturbed to increase irregularity.
- Several replacement rules can be specified, and at each step one replacement rule, chosen at random, is applied.
- Rather than replacing line segments, we might instead look at the trajectory of a particle as we define replacement rules for its position.

Although it is not a very evocative word, the term *fractal* is used to describe the class of shapes that are in some sense rough, but that have a similar appearance at arbitrary scales of magnification. The word derives from the idea that the

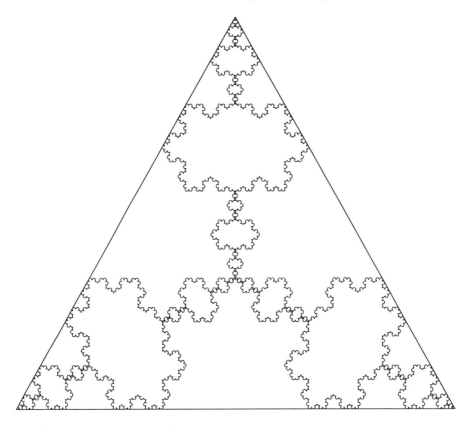

Figure 1.18. A variation with the replacement rule pointing inward.

"dimension" of a shape, which is easy to determine, and a whole number for parametrically or algebraically defined objects, can be extended to a fractional quantity for the kind of shapes described above.[4]

While fractals are of great interest, they are but an example of models that can be defined *procedurally*: the ideal von Koch curves contain indefinite replacement at arbitrarily-small scales, but in practice the process is terminated at a level that is appropriate for the application. The specification of the curve, and its generation to any level of detail desired, is best captured by an iterative process. Many other kinds of curves can also be described in this manner, including polynomials.

---

4. For further reading, consult H.-O. Peitgen and D. Saupe (editors), *The Science of Fractal Images*, Springer-Verlag, New York, 1988; see also B.B. Mandelbrot, *The Fractal Geometry of Nature*, W.H. Freeman and Co., New York, 1983 (a revision of the 1977 book). An excellent introduction to the efficient computation of randomly-generated curves and surfaces is A. Fournier, D. Fussell, and L. Carpenter, "Computer rendering of stochastic models", *Communications of the ACM* *25*, 6 (June 1982), pp371-384.

```
#      Replace the sequence of line segments in "current" using the
#      replacement rule given by "rule".  Notice that "rule" is
#      the name of a Maple procedure.
Replace := proc(current,rule)
     local n,i,new;
     n := nops(current);
     new := NULL;
     for i from 1 to n-1 do
          new := new, rule(current[i],current[i+1]);
     od;
     [new];
end:


#      Replacement rule for von Koch curve.
RepVonKoch := proc(P,Q)
     local P1, Q1, C;
     P1 := Lirp(P,Q,1/3);
     Q1 := Lirp(P,Q,2/3);
     C  := Rotate(Q1,P1,Pi/3);
     P,P1,C,Q1,Q;
end:


#      Rotate P=[px,py] about Q=[qx,qy] through angle ang.
Rotate := proc( P, Q, ang)
     local s, c, xx,x, y;
     s := evalhf(sin(ang));
     c := evalhf(cos(ang));
     x := P[1]-Q[1];
     y := P[2]-Q[2];
     xx:=  (c*x - s*y + Q[1]);
     y :=  (s*x + c*y + Q[2]);
     [xx,y];
end:


#      Return point that is the fraction t between P and Q.
#      "Lirp" stands for "linear interpolation".
Lirp := proc(P,Q, t)
     [P[1] + t*(Q[1]-P[1]), P[2] + t*(Q[2]-P[2])]:
end:


#      Some possible initial shapes.
Line         := [[0,0], [1,0]]:
InitVonKoch  := [[0,0], [1,0], Rotate([2,0],[1,0], Pi/3),[2,0], [3,0]]:
InitVonKoch1 := [[0,0], [1,0], Rotate([1,0],[0,0],-Pi/3),[0,0]]:
InitVonKoch2 := [[0,0], [1,0], Rotate([1,0],[0,0], Pi/3),[0,0]]:
```

Figure 1.19. Code to implement replacement rules for sequences of line segments.

1.5. Discretisation and Computation of Functions

So far, we have considered the representation of functions defined over the real numbers. We have briefly touched on the issue of their rendering in terms of line segments, which is a simple example of converting ideal mathematical functions to a lower-accuracy representation. In some sense, our conversion above is still of high precision, because the endpoints of line segments making up our approximate representation are still defined over $\mathbf{R}^3$. When dealing with digital computation, however, we must often resort to the *approximation* of functions. Indeed it is often the case that a *discrete* representation of functions is required, meaning that only finite sets of values are available to us for our representations. Two examples of such discrete sets are the *floating point numbers* and *finite prefixes* of the integers, namely $\{-M, -M+1, \cdots, -3, -2, -1, 0, 1, 2, 3, \cdots, N\}$, where M and N correspond to the smallest and largest whole numbers representable with that definition of the type **int**. In both cases, the particular sets of numbers available will depend on the computer on which we run our programs. Compared to the reals (or indeed any infinite set), the set of values available in a computer is very small indeed. We shall discuss floating-point representations in Chapter 5.

In symbolic computation systems, it is occasionally possible and usually desirable to work with implicit mathematical definitions of objects. Sometimes properties such as the length, area, or volume of an object can be derived algebraically. However, in most cases, numerical computation must be employed to extract quantitative information from mathematical objects. This process necessarily involves *approximation*, which arises in several ways. For example:

- The computer representation of real-valued quantities is rarely exact.
- Many functions are difficult to compute and are therefore represented by simpler, approximate functions.
- The domains of functions may be intractably large and are consequently truncated.
- A mathematical function may be highly variable, discontinuous, or otherwise badly behaved over a certain domain, and it may be replaced by a better behaved approximation.
- A finite representation of the function may be required. For example: a compact disk contains a finite set of integral samples from a real-valued musical source; an image from computer graphics is represented in terms of a finite set of picture elements on a display screen.

The satisfactory approximation of a large set of mathematical functions discretely is part of the foundation for our science and technology, whether computer-based or not. To illustrate the process of approximating ideal mathematical objects by concrete computational objects, we shall deal with the discrete depiction of two kinds of parametrically defined functions: line segments and circles. These functions are very simple mathematically, but the amount of thought that has gone into their discrete representation is formidable. In both cases, we shall see that issues of robustness and efficiency drive us toward computational implementations that deviate substantially in form from the original definition.

Our discussion of line discretisation will introduce an approach to program development called *program transformation*, a paradigm that can be applied to a wide variety of programs. The approach we shall take is one on which many numerical methods are based: we approximate the behaviour of a function at the next instant by looking at its current behaviour and its derivative at that point. Furthermore, when we try to render circles, we shall see a good example of how naive approximations can accumulate error, leading to catastrophic results.

1.5.1. Line Segments and Circles

Perhaps the most obvious form of discretisation arises when functions (or equivalently, geometric objects) must be rendered onto a display screen. In this case, we must transform functions defined over $\mathbf{R}^n$ into sets of ordered pairs (i,j), where i and j are natural numbers, the ranges of which are within the resolution of the display device. Each pair represents the position of a pixel on a display screen. Most displays on current bit-mapped workstations have a *resolution* of roughly $1\,024 \times 1\,024$ pixels. This is an extremely small set of numbers with which to play. Before dealing with the issue of discretising line segments, we need some background on pixels and images.

Mathematically speaking, a point $P \in \mathbf{R}^n$ or a line segment $L(P_0, P_1)$ has no area. On the other hand, when viewed as pixels, a point or line segment should be thought of as occupying actual area on the screen, for otherwise it would be invisible. In fact, for reasons of visibility, on many display devices such as plotters, a command to display a ''point'' can result in the display of a fairly large cross, rectangle, or bullet. A *pixel* display primitive on a two-dimensional grid has an address P, and a value v. We assume the following two operations are available:

$$v = \text{GetPixel}(P),$$
$$\text{PutPixel}(P, v),$$

where $P \in \mathbf{N} \times \mathbf{N}$ is an integral grid point, and $v \in \mathbf{C}$ is a colour value. A pixel should also be thought of as occupying area on a display screen, and it thus has a shape as well as an address and value. It differs from ordinary memory in this regard (and in many other ways, although in our case it is reasonable to think of pixels as memory locations and images as two-dimensional arrays). In our case, we shall assume a pixel is a unit square with an integral centre. Pixels, images, and raster graphics in general are discussed further in Appendix A.

This notion of "line drawing" in computer graphics is a slight misnomer, since one usually means "line-segment" drawing. Recall our parametric definition for a line segment L between two points $P_0, P_1 \in \mathbf{R}^n$:

$$
\begin{aligned}
\mathbf{L}(P_0, P_1) &= \{(1-t)P_0 + tP_1 : t \in [0,1]\} \\
&= \{P_0 + t(P_1 - P_0) : t \in [0,1]\}.
\end{aligned}
$$

Both of these equations denote the values of the x and y components of the line segment in terms of a parameter t. If t is allowed to vary over all of $\mathbf{R}$, one gets a parametric line rather than a line segment. As we have seen, if $P_0 = (x_0, y_0)$, and $P_1 = (x_1, y_1) \in \mathbf{R}^2$, then any point (x,y) on the line segment is of the form

$$
\begin{aligned}
x(t) &= x_0 + t(x_1 - x_0), \\
y(t) &= y_0 + t(y_1 - y_0),
\end{aligned}
\tag{4}
$$

for some value of $t \in [0,1]$. For notational brevity, we have written the parametric function in x as $x(t)$ rather than $f_x(t)$, and similarly for y.

Exercise 1.31. Verify that this definition gives the desired line segment.

Exercise 1.32. What is the object defined when t varies over $[0, +\infty]$?

The above definition specifies an unobservable, infinitesimally thin, area-less set. While this may be perfectly acceptable when defining geometric objects, it clearly is not when displaying line segments. We define a line-segment display primitive as:

$$\textbf{LineSegment}(P_0, P_1, v),$$

where $P_0, P_1 \in \mathbf{N}^2$ and $v \in \mathbf{C}$. There are many devices that directly display line segments. These go under a variety of names such as *calligraphic, line-drawing,* or *vector-drawing* display systems, and typically can render high-quality (i.e., seemingly straight) line segments very quickly. For many years, line-drawing systems were the predominant display technology in computer graphics. Some raster systems also contain hardware for drawing line segments very quickly, so that even some architectures whose only real display primitive is the pixel can be thought of as directly supporting a line-segment primitive. Regardless of the device implementing such a primitive, *every* rendering of a line segment necessarily has width, although the quality of rendering can differ widely in terms of perceived straightness, "jagginess," or density.

1.5.1.1. Rendering Line Segments

To depict a function or object O on a raster display screen, we must transform the highly accurate representation of O into a set of pixels that are set to a given colour. This process is called *rendering.*[5] A rendering R is a transformation from

---

5. In a graphics system, this is just one of the many transformations that rendering connotes. We shall be content just with this one operation.

a mathematical representation of O to a display representation, and we can write this operation down as

$$R[O] = \{(i_0, j_0), (i_1, j_1), \cdots, (i_n, j_n)\},$$

where the set of ordered pairs denotes pixels to be coloured by the rendering. Square brackets are used simply to convey the idea that this function has a *discrete* range, and continuous domain.

Our goal is to define a rendering R that maps a class of line segments L that can make up object O to sets of pixels. Specifically, we wish to render line segments $\mathbf{L}(A,B)$, where $A, B \in \mathbf{N}^2$. We shall restrict our attention to line segments in the *first octant* with the first endpoint at the origin. Later we shall see why it is sufficient to do so. For a line segment with one endpoint at the origin to be in the first *octant* means that its slope lies between 0 and 1. Many renderings of $\mathbf{L}$ are possible, but most are based on the following crucial idea. Imagine a particle travelling along the trajectory of a parametric curve. We can estimate the position of the particle in the next instant of time by using the current position together with the derivative of the curve at that position. The following discussion will make this more precise.[6]

Let us suppose that we wish to compute the position of the particle at times $t_0, t_1, \cdots, t_n$. Let the position at time t_i be labelled (x_i, y_i). If f is represented by parametric functions $x(t)$ and $y(t)$ then we define

$$(x_i, y_i) = (x(t_i), y(t_i)), \quad i = 0, 1, \cdots, n.$$

If we happen to know the value of (x_i, y_i) for some i, then we can approximate what happens at the next step $i + 1$ by:

$$
\begin{aligned}
x_{i+1} &\sim x_i + \Delta x(t_i), \\
y_{i+1} &\sim y_i + \Delta y(t_i).
\end{aligned}
\tag{5}
$$

In the case of a line segment we certainly know the values at two such i, since we know that the line rendering must include the endpoints. The two terms involving Δ are related to the derivative of $x(t)$ and $y(t)$ in the following way. Recalling the parametric definition of a line segment in the form given by Eq. 4, we observe that

$$
\begin{aligned}
\frac{d}{dt}x(t) &= x_B - x_A, \\
\frac{d}{dt}y(t) &= y_B - y_A.
\end{aligned}
\tag{6}
$$

This is a very simple set of *differential equations*. Let us assume that the spacing

---

6. The following discussion examines the general principles underlying the use of forward differences based on differential equations. This is overkill for the rendering of line segments, but our discussion demonstrates a general approach to approximating/computing any differentiable function.

between t_i and t_{i+1} is constant and the same for each i. Let Δt be this constant. Then we can write Eq. 6 down in terms of *finite differences*:

$$\Delta x(t_i) = (x_B - x_A)\,\Delta t,$$

$$\Delta y(t_i) = (y_B - y_A)\,\Delta t. \tag{7}$$

We shall further suppose that $x_B > x_A$, and that n, the number of steps in t, is exactly $x_B - x_A$. That is, we wish to sample the line segment for each relevant column in x, and that x_0 corresponds to x_A, x_1 to $x_A + 1$, and so on up to x_n corresponding to x_B. Under this convenient assumption,

$$\Delta t = \frac{1}{x_B - x_A},$$

and therefore Eq. 7 becomes

$$\Delta x(t_i) = \frac{x_B - x_A}{x_B - x_A} = 1,$$

$$\Delta y(t_i) = \frac{y_B - y_A}{x_B - x_A} = m, \tag{8}$$

where $m = \Delta y / \Delta x$ is the slope of the line segment. Folding Eq. 8 back into Eq. 5,

$$x_{i+1} \sim x_i + 1,$$

$$y_{i+1} \sim y_i + m. \tag{9}$$

In the next chapter, we shall see how this approach extends to the fast evaluation of x_{i+1}, y_{i+1} for any parametric polynomial representation of a curve.

Since our derivation applies to many kinds of functions, the best for which we can normally hope is an approximation whose accuracy depends very much on how many samples we take of the function (i.e., how large n is). For a line segment, however, the "approximation" is exact, but we still cannot render it directly because m is an arbitrary rational number, and we need integral co-ordinates for pixels. On the other hand, our parametric functions are now in a form that permits rendering. Many line-rendering schemes have been proposed. The most common one is called the *optimal line (segment) rendering*, which we shall name R_o. We can define our goal formally as follows:

$$R_o[L((0,0), (x_1,y_1))] = \{\text{pixels } (i,j): 0 \le i \le x,\ j = round(i\,m)\}.$$

This will be the basis for an *implementation* of **LineSegment**$((0,0), P_1, v)$. It will be as faithful a representation of Eq. 9 as possible, which is the only reason why we call it "optimal." At the moment there is no evidence at all that it can be *computed* efficiently, however one chooses to define "efficiently." The notation *round*(x) denotes the integer nearest to x. If *int*(x) is the integral component of $x \in \mathbf{R}$, $x \ge 0$, then we define

$$round(x) = int(x + \tfrac{1}{2}).$$

Exercise 1.33. This is not a good definition for negative x. Suggest a better definition.

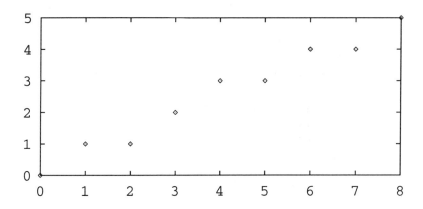

Figure 1.20. Optimal line-segment rendering from (0,0) to (8,5).

Figure 1.20 gives a sample rendering from the point (0,0) to (8,5). The optimal rendering has the following properties for line segments in in the first octant:

- *Exactly* one pixel per column is chosen for each relevant column within by $[0,x]$ of the line segment.
- If pixel (i,j) is chosen in column i, then the next pixel chosen to the right (i.e., column $i+1$) is either row j or $j+1$. That is, the pixel chosen in the next column is either to the right, or both to the right and up one unit.

A simple and inefficient implementation of R_o can be found in Figure 1.21.

We have not yet discussed floating-point arithmetic, but the only relevant issue at play here is that the arithmetic using the **real** type in Turing (or **float/double** in C) is often much slower than integer arithmetic.[7]

Exercise 1.34. Prove that R_o has the two properties claimed above.

Exercise 1.35. Implement a Turing or C program that takes as input a sequence of pairs of natural numbers and colours, and uses the code from Figure 1.21 to render line segments rooted at the origin.

---

7. On computers equipped with special floating-point processors this is not true. However, line-rendering algorithms are often implemented in microcode or hardware that do not have access to floating-point processors.

```
% Draw line segment from (0,0) to (x,y), y ≤ x.
% Thus x = Δx and y = Δy

proc DrawLine(x,y: int; c: colour)
    var yt, m: real
    var yi: int
    if x=0 then error
    m   := y / x
    for xi: 0..x
        yt := m*xi
        yi := floor(yt + ½)        % round(yt)
        PutPixel(xi,yi,c)
    end for
end DrawLine
```

Figure 1.21. A brute-force optimal line renderer. Note that in Turing, unlike C, the expression i/j for integer i,j returns a result of type **real**.

We shall successively refine the program in Figure 1.21 until we get to something more efficient and more stable. We shall get to an equivalent line-rendering algorithm that requires no floating-point arithmetic at all. Notice that, just as when we discussed the rendering of polynomials, we again have been provided with a mathematical specification, namely R_o, of our task. Each of the programs in our sequence of transformations will be a realisation of the specification.

The refinement technique that we shall use is sometimes called *program transformation*, in that we are actually refining a working implementation to make it more efficient.[8] First, we note that we do not have to multiply to get a correct value of **yt**: we can simply replace the multiplication by an addition. This is illustrated in Figure 1.22. Next, we note that we can avoid costly extra operations by starting our value of **yt** off at ½ rather than 0. A *round* operation now becomes a *floor*, or truncation operation. See Figure 1.23.

So far, we have only made some trivial optimisations to illustrate the use of program transformation. We must dig a little deeper to get better results. We can split **ys** into two pieces: one contains the integral part (**ysi**) and the other contains a fraction between 0 and 1 (**ysf**). Thus **ys = ysi + ysf**. This means we have to be slightly careful about how we increment by **m**. Think of **ysi** and **ysf** as being two buckets and that a container with **m** milliliters of water is to be added to the one-liter **ysf** bucket. If we were to add water that

---

8. This treatment is inspired by R.F. Sproull, "Using program transformations to derive line-drawing algorithms," *ACM Transactions on Graphics 1*, 4 (Oct., 1982), pp. 257-273.

```
proc DrawLine(x,y: int; c: colour)
    var yt, m: real
    var yi: int
    if x=0 then error
    yt := 0
    m   := y / x
    for xi: 0..x
        yi := floor(yt + ½)
        PutPixel(xi,yi,c)
        yt += m
    end for
end DrawLine
```

Figure 1.22. Transformation 0: remove multiplication.

```
proc DrawLine(x,y: int; c: colour)
    var ys, m: real
    var yi: int
    if x=0 then error
    ys := ½
    m   := y / x
    for xi: 0..x
        yi := floor(ys)
        PutPixel(xi,yi,c)
        ys += m
    end for
end DrawLine
```

Figure 1.23. Transformation 1: substitute **ys** for **yt** + ½.

would overflow the **ysf** bucket, then fill the **ysf** bucket to the top (meaning that it would contain one liter of water), pour the **ysf** bucket into the **ysi** bucket, thereby adding one liter to the much larger **ysi** bucket, and fill the now empty **ysf** bucket with the leftover water from the **m** milliliters we originally had to pour. See Figure 1.24.

The most important transformation is also the most difficult to understand. The idea is that we would like to replace the floating-point variable **ysf** by an integer. Let us examine the behaviour of **ysf** to see how this can be done. First note that **ysf** has always been added to by **y/x** or by $(y/x) - 1$, which is equal to $(y - x)/x$. That is, **ysf** has an **x** in its denominator. Multiplying **ysf** by **x**, which we know not to be zero, removes this part of the denominator. Furthermore, **ysf** is initially ½. Multiplying by 2 removes the 2 from the

```
proc DrawLine(x,y: int; c: colour)
    var ysf, m, m̄: real
    xi, ysi: int
    if x=0 then error
    m   := y / x
    m̄   := m − 1
    ysi := 0
    ysf := ½

    for xi: 0..x
        PutPixel(xi,ysi,c)
        if (ysf+m ≥ 1) then          % ysf overflow?
            ysi += 1
            ysf += m̄
        else
            ysf += m
        end if
    end for
end DrawLine
```

Figure 1.24. Transformation 2: split **ys** into integral (**ysi**) and fractional parts
(**ysf**).

denominator. Therefore $2(\textbf{ysf})\textbf{x}$ is an integer. By definition, $\textbf{ysf} \in [0,1]$. Thus the quantity $(\textbf{ysf} - 1) \in [-1,0]$, and by the same argument as above, $2(\textbf{ysf} - 1)\textbf{x}$ is also integral. The reason for subtracting one from **ysf** will become apparent momentarily. It is also convenient to add $2\textbf{y}$ to this whole expression, giving it the name **r**:

$$r = 2\textbf{y} + 2(\textbf{ysf} - 1)\textbf{x}. \tag{10}$$

Since **y** is an integer, **r** must also be. We shall use **r** in place of **ysf** in our code, for it happens to have some wonderful properties in addition to it being an integral quantity. Note that while all of our other quantities were positive, **r** can in fact be negative.

When is $\textbf{r} \geq 0$? This is easy, for

$$\textbf{r} \geq 0 \;\Rightarrow\; 2\textbf{y} + 2(\textbf{ysf} - 1)\textbf{x} \geq 0$$

$$\Rightarrow\; \frac{2\textbf{y}}{2\textbf{x}} + \textbf{ysf} - 1 \geq 0$$

$$\Rightarrow\; \textbf{ysf} + \textbf{y}/\textbf{x} \geq 1.$$

This is precisely the test in the **if** statement above. Magic! This expression **r** is an integral representation of the vertical error from the centre of the chosen pixel to the ideal line segment. In effect, it tracks when to go "up and to the right" as opposed to just "to the right." To transform our algorithm to use $\mathbf{r}$, we do the following:

- To initialise, we must preserve the fact that $\mathbf{ysf} = \frac{1}{2}$. From Eq. 10, this implies $\mathbf{r}$ is initialised to $2\mathbf{y} - \mathbf{x}$.
- The statement $\mathbf{ysf} = \mathbf{ysf} + \mathbf{m}$ means that $\mathbf{r} = \mathbf{r} + [2\mathbf{y}]$.
- The statement $\mathbf{ysf} = \mathbf{ysf} + \bar{\mathbf{m}}$ means that $\mathbf{r} = \mathbf{r} - [2\mathbf{x} - 2\mathbf{y}]$. The notation $[z]$ just means that z is a constant and integral.

This comes directly from substituting the new value of $\mathbf{ysf}$ into the expression for $\mathbf{r}$. Putting it all together gives us *Bresenham's algorithm* (Figure 1.25).

```
proc DrawLine(x,y: int; c: colour)
    var r, ysi: int
    const inc1 := 2y-2x
    const inc2 := 2y

    if x=0 then error
    ysi := 0
    r    := 2y - x

    for xi: 0..x
        PutPixel(xi,ysi,c)
        if (r ≥ 0) then              % ysf overflow?
            ysi += 1
            r += inc1
        else
            r += inc2
        end if
    end for
end DrawLine
```

Figure 1.25. *Bresenham's Algorithm.* Transformation 3: use $\mathbf{r} = 2\mathbf{y} + 2(\mathbf{ysf}-1)\mathbf{x}$ in place of $\mathbf{ysf}$.

Some remarks are in order. First, the implementation is expressed purely in integer arithmetic. Second, no multiplications are required (multiplication by 2 is a left shift—see Chapter 5). Third, it is fast and stable. Fourth, it is much harder to understand than the first algorithm presented.

Exercise 1.36. Implement Bresenham's algorithm in the program you wrote above, and compare the time to render many line segments using the brute-force routine to Bresenham's algorithm. The timing need not be sophisticated. It suffices to write a loop that draws many line segments and time the results using a watch.

Exercise 1.37. So far, your program can only draw line segments from the origin to a value (**x,y**), where $0 \le y \le x$. Extend **DrawLine(x,y)** to **DrawLine(x0,y0,x1,y1)**, where **x0,x1,y0,y1** are all natural numbers. Hints:

- To draw from, say, (8,5) to (0,0), you can draw from (0,0) to (8,5); in other words, reverse the endpoints in some cases.
- To draw from, say, (2,1) to (8,5), pretend you are drawing from (0,0) to (8-2,5-1) and add 2 to the **x** values and 1 to the **y** values.
- To draw line segments with slopes outside of the range [0,1], you will have to interchange **x** and **y** values and/or negate them. That is, for each octant, you may find that all **x**s become **y**s.

Work through each of these possibilities one by one to get an efficient line-rendering algorithm. State-of-the art line-drawing algorithms are not much different from this (except that they are often implemented in machine language or in hardware).

1.5.1.2. Rendering Circles

Since we were successful in converting the problem of rendering line segments into an iterative algorithm, perhaps the same approach will work for the rendering of circles. If we can render circles centred at the origin, we can render them all (why?), so we restrict ourselves to the problem of rendering the following parametric equations

$$x(u) = r\cos u,$$
$$y(u) = r\sin u,$$

where $0 \le u < 2\pi$, and r is the radius of the circle in pixel units. Then

$$\frac{d}{du}x(u) = -r\sin u,$$
$$\frac{d}{du}y(u) = r\cos u,$$

and therefore

$$dx = -r\sin u\, du,$$
$$dy = r\cos u\, du.$$

If as before we choose a (small) uniform spacing Δu for $u \in [0,2\pi]$, so that $u_i = i\Delta u$, then we can write down difference equations for computing point (x_{i+1},y_{i+1}):

$$x_{i+1} \sim x_i - r\sin(u_i)\Delta u = x_i - y_i\Delta u,$$
$$y_{i+1} \sim y_i + r\cos(u_i)\Delta u = y_i + x_i\Delta u. \tag{11}$$

In practice, Δu is chosen to be 2^{-m}, where m is such that

$$2^{m-1} \le r \le 2^m.$$

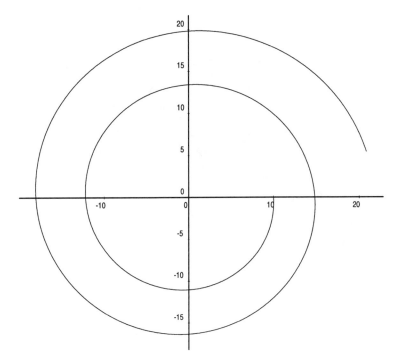

Figure 1.26. A circle?

In this case, the multiplications can be reduced to shift operations. The theory seems reasonable enough, but the spiral in Figure 1.26 is the result. The spiral has been scaled greatly so that it is obvious. Normally, successive arcs of the spiral would only be separated by one or two pixels. What happened? It is not at all obvious, but one way of looking at the problem is as follows. Consider the tangent line T at a point P on a circle as in Figure 1.27. If we follow the tangent line T in either direction then we are being taken *away* from the circle. Even with extremely small steps, after plotting many points, the points will inevitably be pulled off the circle.

There is a "fix," but it is neither entirely satisfactory nor easy to explain:[9]

$$x_{i+1} = x_i - y_i \Delta u,$$
$$y_{i+1} = y_i + x_{i+1} \Delta u. \tag{12}$$

In other words, we use x_{i+1} instead of x_i to compute y_{i+1}! A precise explanation of

---

9. See also W. Newman and R. Sproull, *Principles of Interactive Computer Graphics*, 2nd edition, McGraw-Hill, NY, 1979.

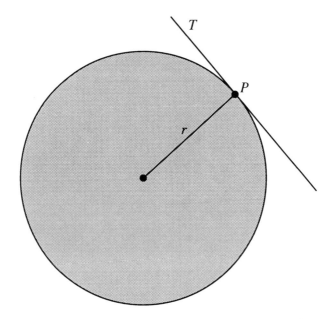

Figure 1.27. Circle with radius r, and tangent line T at point P on the circle.

this requires some linear algebra and takes us too far afield. The gist of the explanation is this. Write down matrix representations for the operations in Equations 11 and 12 that take (x_i, y_i) to (x_{i+1}, y_{i+1}). The "drift" of these computed points from the circle can be estimated by looking at the *determinant* of each matrix. The determinant corresponding to the operations in Eq. 11 is $1 + \Delta u^2$ while the determinant from Eq. 12 is 1. In the first case, a small but consistent distortion of space is being applied, while the second transformation essentially corrects for this distortion. The result can be seen in Figure 1.28. We have paid a small price for this operation: as it happens we have actually rendered a slightly eccentric circle, or in other words, an ellipse. The eccentricity of the ellipse is directly related to Δu, and has been exaggerated in Figure 1.28 for illustration. At normal scales for circles of greater than a few pixels in radius, the result is invisible.

A great deal more research has been done on the rendering of circles, and in fact the algorithm we presented here is not state-of-the-art. It was chosen to illustrate the problems that can arise when one is not careful in taking mathematical models to implementations.

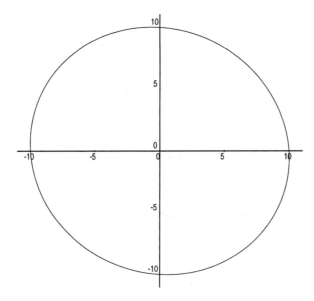

Figure 1.28. A better circle. Note the slight eccentricity (which has been exaggerated for display purposes).

Appendix A: Raster Graphics Fundamentals

This brief appendix outlines some additional background information on the composition of raster graphics systems.

1.A.1. Pixels

Over the 1970s and 1980s, as the price of computer memory decreased, pixel-based or *raster* display devices began to proliferate to the extent that they are now by far the most popular display technology in computer graphics. The basic display primitive in a raster system is the *picture element*, which historically has been abbreviated to *pixel*. A pixel can be thought of as having two components: a shape and a value. Part of the value of a pixel is used to define its colour, which is assumed to be a constant across the region of the image the pixel covers.

Much of the discussion in this section is pertinent to hardcopy display devices as well as raster display systems. Many hardcopy devices now allow individual display "spots" to be manipulated, although it is generally not possible to read a value back from a hardcopy device. In this sense, such devices are "write-only" devices.

1.A.1.1. Pixel Shapes and Rasters

A single pixel on its own is hardly interesting. There have to be many of them arranged in particular ways in order to devise effective and efficient rendering techniques. A *raster* is thought of as covering a desired viewing area in the viewing plane, and consists of a set of pixels arranged in a grid so that every point in the viewing area is covered by at least one pixel, and such that each pixel can be uniquely addressed by an ordered pair $(i,j) \in \mathbf{N}^2$. The pixels in a raster are usually non-overlapping, and we shall assume that in fact pixels are unit squares.

As a data type, the elements of a raster are essentially addressed as in a regular 2-D array, and it can be defined as follows:

> **raster**:
> > Xsize: **integer** = X_RESOLUTION
> > Ysize: **integer** = Y_RESOLUTION
> > I: **array**[0..Xsize − 1, 0..Ysize − 1] **of** pixel

A pixel data type might be defined as:

> **pixel**
> > Shape: connected subset of $\mathbf{R}^2$
> > Value: **pixelValue**

Technically a pixel shape can be any connected set, but is generally a triangle, rectangle, or hexagon. Alternative names for a raster are *image*, *image store*, or *frame buffer*. If A is a raster, then the elements of A are denoted by *A.Xsize*, *A.Ysize*, and *A.I* [i,j], $0 \le i < A.Xsize$, $0 \le j < A.Ysize$. Often, the dot notation will be omitted if the raster in question is obvious from the context. A raster has *resolution Xsize* horizontally, and *Ysize* vertically. The number of pixels in the raster is thus *Xsize* × *Ysize*. We shall assume that position (0,0) denotes the bottom-left pixel of the display, that values in *x* increase to the right, and that values in *y* increase upward.

1.A.1.2. Pixel Values

In addition to a shape, each pixel in a raster has a value. At the very least, each pixel must have a specified colour, but more information is often present.

There are four common possibilities for the meaning of pixel values and the amount of information required to store them. Naturally, the variety of possible pixel operations will depend on the class of pixel values on which they work. A pixel value may be:

- a bilevel or one-bit value. Each pixel value contains a single bit of information, typically denoting whether that pixel is to be given one of two colours, although sometimes it denotes other things, such as a mask.

- a grey value. Each pixel value contains several bits of information capable of representing a discrete graduation of luminance (grey) values from black up to white.

- a colour value or a tuple of grey values. Each pixel contains three (or more) fields of information, which represents a discrete co-ordinate in an underlying colour space.
- an indirect pixel value specification. The value is to be used as an index into a look-up table to determine the colour of the pixel to be displayed.

While on the one hand, a frame buffer can be thought of as simply a two-dimensional array of memory, another modality of it is that its contents are somehow automatically mapped to a display device. In practice, a display device reads or *scans* the frame buffer at a fixed frequency. Common scanning frequencies are 25Hz, 30Hz, 50Hz, and 60Hz, depending on display technology and the frequency of electrical alternating current. Typically, values from the frame buffer are scanned one pixel at a time, row-wise. Let A be a raster. Then Figure 1.29 simulates the scanning activity.

for $j = 0$ to $A.Ysize-1$
    for $i = 0$ to $A.Xsize-1$
        Scan($A.I[i,j]$)

Figure 1.29. Simulation of a raster scan.

We assume the procedure Scan() implements the hardware function of scanning a particular pixel onto a display. The code in Figure 1.29 is a simplification, since in some cases only alternating rows are taken at any given time (this is called *interlacing*).

Exercise 1.38. If A is a picture of resolution $1\,024 \times 1\,024$, what is the maximum time that can be taken by Scan() to be able to display a new picture every 1/60 seconds (assume loop control is negligible).

In case it is not obvious by now, the term *raster graphics* refers to the body of concepts and techniques in computer graphics that have been developed specifically for displaying on raster devices. Several aspects of computer graphics are unique to raster graphics, such as pixel-based rendering algorithms, illumination, and shading. One of the most important algorithms in raster graphics is the efficient rendering of line segments and arcs. The first is discussed above, and the second will be discussed later.

Appendix B: Simple Maple Examples

*A **Sphere**.* The following code creates a three-dimensional plot of a sphere of radius 1.

```
x := sin(phi)*sin(theta):
y := sin(phi)*cos(theta):
z := cos(phi):

plot3d( [x, y, z],
     theta=0..2*Pi, phi=-Pi..Pi,
     title=`Sphere of radius 1 defined parametrically` );
```

A Spiral. This creates the spiral in Figure 1.6.

```
with(plots):
spacecurve( [cos(t), sin(t), t], t=0..6*Pi, numpoints=100);
```

***Circle Drawing*.** The following code simulates the two circle drawing algorithms discussed above.

```
ceiling := proc(x)
    local y;
    if (frac(x) > 0) then
        trunc(x)+1;
    else
        trunc(x);
    fi;
end:
#
#    Take the log(x) base 2, given built-in natural log.
#
log2 := proc(x)
    evalf(log(x)/log(2));
end:
```

```
#
#    This routine tries to draw a circle using forward differences.
#    It plots a spiral instead.
#
circle1 := proc(r)
    local x, y, eps, n, m, pairs;

    x[1] := r;                # one point comes for free
    y[1] := 0;
    n    := ceiling(log2(r));
    eps  := 2^(-n+1);
    m    := round(evalf(Pi*r*r));

    for i from 1 to m do
        x[i+1] := evalf(x[i] + eps*y[i]);
        y[i+1] := evalf(y[i] - eps*x[i]);
    od;
    pairs := [x[j],y[j]] $ j=1..m;
    plot([pairs]);
end:


#
#    This routine corrects the outward spiral.
#
circle := proc(r)
    local x, y, eps, n, m, pairs;

    x[1] := r;                # one point comes for free
    y[1] := 0;
    n    := ceiling(log2(r));
    eps  := 2^(-n+1);
    m    := round(evalf(Pi*r*r));

    for i from 1 to m do
        x[i+1] := evalf(x[i] + eps*y[i]);
        y[i+1] := evalf(y[i] - eps*x[i+1]);
    od;
    pairs := [x[j],y[j]] $ j=1..m;
    plot([pairs],style=LINE);
end:
```

Appendix C: Matrix Representations

Matrix representations of linear equations, vectors, and linear transformations form the core of linear algebra. In this book, the use of matrix representations is largely for convenient notational and computational expressions. That said, it is important to give a brief overview of their use.

Vectors (and in some cases, points) can be represented in matrix algebra as rows or columns of values, enclosed by square braces. For example, we can write the vector $\mathbf{u} = (u_1, u_2, u_3)$ as either a *row vector*

$$\mathbf{u} = [\, u_1 \; u_2 \; u_3 \,],$$

or as a *column vector*

$$\mathbf{u} = \begin{bmatrix} u_1 \\ u_2 \\ u_3 \end{bmatrix}.$$

A *matrix* is a rectangular grid of values consisting of n rows and m columns; such a matrix is said to be an $n \times m$ matrix. Most of the matrices we shall be working will be square, i.e., $n = m$. A row vector is notationally just a $1 \times m$ matrix, and a column vector is an $n \times 1$ matrix. We can write a 4×3 matrix M as

$$M = \begin{bmatrix} a & b & c \\ d & e & f \\ h & i & j \\ k & l & m \end{bmatrix}.$$

An individual element of a matrix M can be addressed as M_{ij}, meaning the element that is both in row i and column j. We can rewrite M above as

$$M = \begin{bmatrix} M_{11} & M_{12} & M_{13} \\ M_{21} & M_{22} & M_{23} \\ M_{31} & M_{32} & M_{33} \\ M_{41} & M_{42} & M_{43} \end{bmatrix}.$$

The elements of a matrix are not required to be scalar values: they could be vectors or other matrices, though sometimes operations on matrices with non-scalar elements may not make sense.

Matrices can be multiplied together to create new matrices. Its definition is more intricate than scalar multiplication. If A is an $n \times p$ matrix and B is a $p \times m$

matrix, then their multiplication

$$C = A B$$

gives an $n \times m$ matrix, and is defined for each element of C as follows:

$$C_{ij} = \textbf{row } i \textit{ of } A \times \textbf{column } j \textit{ of } B$$

$$= \sum_{k=1}^{p} A_{ik} B_{kj}.$$

Observe that the number of columns of A must be equal to the number of rows of B. As a special case, if $\textbf{u}$ is an n-element row vector (which is represented as a $1 \times n$ matrix) and if $\textbf{v}$ is an n-element column vector (or an $n \times 1$ matrix), then their product is a one-element matrix. In other words, the product is a scalar. This is in fact the *scalar*, *inner*, or *dot* product of $\textbf{u}$ and $\textbf{v}$. Normally a different notation is used to denote scalar products. If $\textbf{a} = (a_1, a_2, a_3)$ and $\textbf{b} = (b_1, b_2, b_3)$ are two vectors, then the dot product of $\textbf{a}$ and $\textbf{b}$, denoted $\textbf{a}\cdot\textbf{b}$, is defined as

$$\textbf{a}\cdot\textbf{b} = \sum_{i=1}^{3} a_i b_i$$

$$= a_1 b_1 + a_2 b_2 + a_3 b_3.$$

The summation form illustrates that dot products extend to vectors of arbitrary dimension. A dot product has many useful properties such as the fact that if $\textbf{a}\cdot\textbf{b} = 0$ then $\textbf{a}$ and $\textbf{b}$ are perpendicular. Another important property is that for vectors $\textbf{a}, \textbf{b}, \textbf{c}$,

$$\textbf{a}\cdot(\textbf{b} + \textbf{c}) = (\textbf{a}\cdot\textbf{b}) + (\textbf{a}\cdot\textbf{c}).$$

Furthermore, if $t \in \textbf{R}$, then

$$\textbf{a}\cdot(t\,\textbf{b}) = t\,(\textbf{a}\cdot\textbf{b}).$$

In other words, the dot product distributes over vector addition and *scalar* multiplication.

Exercise 1.39. Verify these last two properties of the dot product from its definition above.

Unlike regular scalar multiplication, matrix multiplication does not commute, meaning that in general $AB \neq BA$; the inner product, which is one kind of vector multiplication, *is* commutative. Matrix multiplication is, however, *associative*, meaning that the order in which we perform more than one multiplication is irrelevant. That is,

$$A(BC) = (AB)C.$$

The *transpose* A^+ of a matrix A is defined by interchanging rows and columns. That is,

$$A_{ij}^+ = A_{ji}.$$

Under this operation, a row vector transposes into a column vector and vice versa.

The *identity* matrix is a square, $n \times n$ matrix with each element I_{ij} equal to 1 when $i = j$ and 0 otherwise. For example, the 3×3 identity matrix is

$$I = \begin{bmatrix} 1 & 0 & 0 \\ 0 & 1 & 0 \\ 0 & 0 & 1 \end{bmatrix}.$$

A matrix structure together with transposition and multiplication is remarkably expressive. For example, matrices can be used to represent geometric transformations of space (such as rotations and scalings). If A and B are two rotations about, say, the x axis, then their multiplication AB represents the composite rotation about the x axis. In a different setting, suppose we have the following set of equations that must be simultaneously satisfied in the unknowns x, y, and z:

$$Ax + By + Cz = B_1,$$

$$Dx + Ey + Fz = B_2,$$

$$Gx + Hy + Iz = B_3,$$

where all the other terms are given. Then we can represent this *linear system of equations* compactly as

$$A\mathbf{x} = \mathbf{b}.$$

The column vector $\mathbf{x}$ contains the variables

$$\mathbf{x} = \begin{bmatrix} x \\ y \\ z \end{bmatrix}.$$

The column vector $\mathbf{b}$ contains the known values:

$$\mathbf{b} = \begin{bmatrix} B_1 \\ B_2 \\ B_3 \end{bmatrix}.$$

The 3×3 matrix A contains the coefficients of the equations

Exercise 1.40. Write out the elements of A.

This system is *solved* for **x** by finding a matrix A^{-1}, the *inverse of A*, such that

$$A^{-1}A = I.$$

If we can find such an inverse, we have a solution for **x**, since

$$A^{-1}\left(A\mathbf{x}\right) = A^{-1}\mathbf{b},$$
$$\left(A^{-1}A\right)\mathbf{x} = A^{-1}\mathbf{b},$$
$$I\mathbf{x} = A^{-1}\mathbf{b},$$

meaning that

$$\mathbf{x} = A^{-1}\mathbf{b}.$$

In linear algebra, a powerful technique called *Gaussian elimination* is used to solve systems of linear equations. For the purposes of this book, we usually just need to be aware of how to multiply matrices and how to encode values as row and column vectors. We shall let Maple invert matrices for us.

Example 1.6. We have written a polynomial of degree n over parameter t as

$$p(t) = a_0 + a_1 t + a_2 t^2 + \cdots + a_n t^n.$$

It can be written in matrix form as

$$p(t) = [t^n \ t^{n-1} \ \cdots \ t^1 \ t^0] \begin{bmatrix} a_n \\ a_{n-1} \\ \vdots \\ a_0 \end{bmatrix}.$$

Of course the ordering of the elements is unimportant—making sure the correct elements are multiplied together is. We shall see in the next chapter that a matrix representation for polynomials is very convenient. In fact we will be solving small systems of linear equations to get different representations of polynomials.

Supplementary Exercises

Exercise 1.41. Consider the polynomial

$$p(x) = x^2 - x + 1.$$

Reparameterise $p(x)$ from the interval $x \in [-1,2]$ to an identically behaving polynomial $\hat{p}(t)$ on the interval $t \in [0,1]$.

Exercise 1.42. Write a Maple procedure `reparam(p,x,[a,b],t)` that takes a polynomial `p` in indeterminate `x`, a domain `[a,b]`, and a free variable `t` as arguments. The routine should return a new polynomial $\hat{p}(t)$ that is the reparameterisation of the polynomial $p(x)$ over $[a,b]$ onto the domain $[0,1]$. Suppose, for example, I wish to reparameterise the polynomial $x^2 - 1$ from the domain $[-2,2]$ to $[0,1]$. Then your routine should output a simplified polynomial as follows:

```
> p := x^2 -1:
> reparam(p, x, [-2,2], t);
                        2
              16 t  - 16 t + 3
```

Exercise 1.43. Let $p(t)$ be a polynomial that is parameterised on $[0,1]$. Show how to reparameterise $p(t)$ to an identical polynomial $\bar{p}(u)$ with $u \in [a,b]$, for arbitrary $a,b \in \mathbf{R}$, $a < b$. Is the degree of p the same as the degree of $\bar{p}$? Explain.

Exercise 1.44. Sketch an (x,y) plot of each of the following parametrically defined functions given by $F(t) = (x(t),y(t))$. In each case, $t \in [-1,1]$.

(a) $F(t) = (t^2, t)$.

(b) $F(t) = (t^2, -t^2)$.

(c) $F(t) = (t^2, t^3)$.

Exercise 1.45. Consider the parametric definition of a curve $C(t) = (x(t), y(t))$ given by

$$x(t) = \sqrt{e^t},$$
$$y(t) = e^t,$$

where $t \in [-10,10]$. There is a simpler parameterisation, $\bar{C}(u)$, of the object denoted by $C(t)$ in which u is some function of t. Derive this reparameterisation including the effect on the domain of the parameter, and plot the result. What object is denoted by C and $\bar{C}$?

Exercise 1.46. We shall analyse the error behaviour of the optimal line rendering. Define ε_i to be the *vertical error* made in pixel i by the optimal line rendering, where $0 \le i \le x$. That is, for each such i, ε_i is the (signed) difference between the real y value of the line segment at pixel i and the integral value of y at (i,j) chosen by the rendering.

(a) Write down a simple mathematical expression for ε_i and implement it in Maple.

(b) This isn't difficult! For a line segment from $(0,0)$ to (x,y), show that

$$|\varepsilon_i| = |\varepsilon_{x-i}|,$$

where $|x|$ denotes the absolute value of x.

Exercise 1.47. Remember our old friend the trochoid? Well, it so happens that the Canadian military might be willing to pay us big bucks if you can tell them interesting things about it and if you can render it quickly. Recall its definition:

$$x(u) = au - r \sin u$$
$$y(u) = r \cos u,$$

where a and r are constants.

(a) Describe and then sketch a plot of the trochoid in the (x,y) plane if a is set to 0, r is non-zero, and $0 \le u < 2\pi$.

(b) Describe and then sketch a plot of the trochoid in the (x,y) plane if r is set to 0, a is non-zero, and $0 \le u < 2\pi$.

(c) Here is a brute-force Turing program that renders the trochoid for non-zero values of a and r.

```
var xn,yn: int
var x,y, u, r, a, du: real
r   := 100                    % for example
a   := 25
du := 0.01
u   := 0
x   := 0
y   := r
loop
      xn := round(x)
      yn := round(y)
      drawdot(xn,yn,1)
      u   += du
      exit when u > 6π
      x   := a*u - r*sin(u)
      y   := r*cos(u)
end loop
```

Transform this implementation into an incremental algorithm so that

$$x_{i+1} := x_i + \frac{dx}{du}\Delta u \,,$$

$$y_{i+1} := y_i + \frac{dy}{du}\Delta u \,.$$

Derive the appropriate derivatives and rewrite them so that *no* sine or cosine function appears anywhere in the result. Update the inner loop in the above code to reflect your derivation.

Chapter 2
Interpolation

The problem of constructing or reconstructing continuous curves from a small set of sample points arises in a great many applications. Creating a curve that passes through these points is called *interpolation*. In this chapter we shall explore some basic issues in interpolation, and we shall contrast global interpolation approaches such as Lagrange interpolation to *piecewise parametric* polynomial interpolation. Our main motivating application will be the simulation of spatial trajectories. As in the previous chapter, we shall also consider a specific computational problem, in this case that of efficiently rendering or evaluating space curves represented as piecewise parametric polynomials.

2.1. A Motivating Problem

We shall return to a problem introduced in the last chapter as motivation for much of the material in this chapter, but we shall make it a little more dramatic. Suppose we have a fly in our room. As part of our long-term plan to monitor the flight patterns of flies, leading to their ultimate extermination, we shall start small by trying to model the trajectory of a single fly over time. At a recent fund-raising auction for the Canadian military, we were able to purchase a robotic *FlyTracker*, which is able to scan the room periodically and to locate the position of a single fly. It then returns a point $(x,y,z) \in \mathbf{R}^3$ and a time t at which it was able to observe the fly. Thus the device produces a sequence

$$(t_0,x_0,y_0,z_0), \ (t_1,x_1,y_1,z_1), \ (t_2,x_2,y_2,z_2), \ \cdots \qquad (1)$$

of values. Think of x, y, and z as denoting left-right, forward-backward, and up-down positions, respectively, as in Figure 2.1, reproduced from Chapter 1.

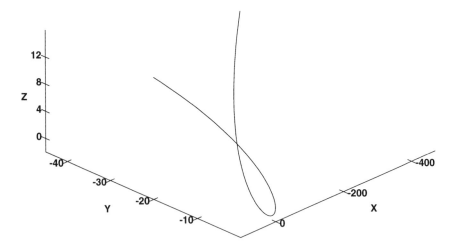

Figure 2.1. The co-ordinate system of the tracker, and a partial trajectory. Notice
that our point of view need not coincide with that of the tracker.

Our goal is to construct a curve that closely approximates the trajectory of the fly
at any time t, not just at the discrete, reported times. Allowing more than one fly
to enter the room may cause the FlyTracker to oscillate so wildly that it may self-
destruct, taking half of our planet with it. Purchases from the military may be
more costly than we think!

    We shall often assume that the spacing between samples is constant. When
this is the case,

$$t_i - t_{i-1} = \Delta t, \tag{2}$$

for all $i > 0$, and moreover,

$$t_i = i \Delta t. \tag{3}$$

Problems like this come up repeatedly in many guises, so our solutions to the
continuous tracking of flies may be usable in other applications. In the last
chapter, we implied that parametric representations may be useful. In this
chapter, we shall see that good solutions to this problem can be found by using
so-called piecewise parametric polynomials. These objects are simpler than they
sound, but first we must examine why the use of polynomials is potentially viable.

2.2. Properties of Polynomials

We have previously defined a polynomial of *degree n* and *order n*+1 as

$$p(t) = a_0 + a_1 t^1 + a_2 t^2 + \cdots + a_n t^n, \tag{4}$$

where $a_i \in \mathbf{R}$, $i = 0, 1, \cdots, n$ and $a_n \neq 0$. As seen in Figure 2.2, depending on its degree, the value of a polynomial wiggles around for awhile, and then tails off to $-\infty$ or $+\infty$. In fact, a polynomial of degree n with real coefficients has exactly n not necessarily distinct zeros. The values of t for which the polynomial is zero are called the *roots* of p. In general, some of these roots may be complex numbers. An important concern is that if we choose to use polynomials to model trajectories we must be careful about the choice of the degree of a polynomial: polynomials of very low degree may not be flexible enough, but those of high degree may be too variable and hard to control.

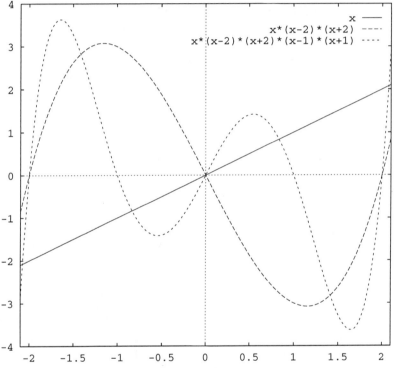

Figure 2.2. Polynomials of degree 1, 3, and 5.

We mentioned earlier that a polynomial has a compact representation: $n+1$ independent pieces of information are sufficient to represent the polynomial. One such set is the coefficients $a_0, a_1, \cdots, a_n$ representing the curve. However, we shall see later that there are other possible representations for a polynomial. Polynomials also have the very nice property of having easily computable derivatives and integrals (for most of the common representations of them).

Exercise 2.1. Using Eq. 4, and retaining its form, what is the derivative and what is the indefinite integral of an arbitrary polynomial of degree n? What are the degrees of the resulting polynomials?

What evidence is there that the polynomials are a sufficiently rich class of functions to allow us to solve our tracking problem? One significant piece of evidence is the *Weierstrass approximation theorem*, which states that for any continuous function, there exists a polynomial (of sufficiently high degree) that approximates the function arbitrarily closely. The proof of this theorem is rather subtle, and it will be faced in all its glory in a mathematical analysis course. Sometimes, mathematicians are satisfied with existence proofs, but computer scientists, who wish to implement practical solutions, need more than the mere fact that a solution exists. In this case, the mathematicians Bernstein, and later Stone, derived proofs of the Weierstrass theorem that show how to *construct* such an approximation, and now we know that many constructions are possible. The topic of approximation will be discussed at length in the next chapter.

While the problem of approximating a function is related to our problem, recall that our primary concern is to come up with a function that also interpolates a set of given sample values. We are adding an extra constraint, and we are taking it on faith that an interpolating function will also be a good approximating function. Unfortunately, this is not always the case, as we shall now see.

2.3. Lagrange Interpolation

We can try to treat our tracking problem as a problem in interpolation in the following way. We shall suppose that our tracker has given us $n+1$ samples

$$(t_0,x_0,y_0,z_0),\ (t_1,x_1,y_1,z_1),\ (t_2,x_2,y_2,z_2),\ \cdots,\ (t_n,x_n,y_n,z_n). \tag{5}$$

We will not assume that the t_i are equally spaced in time, only that $t_i < t_{i+1}$. An appropriate model for the fly's trajectory is a space curve $(x(t),y(t),z(t))$ in time parameter t. To construct a space curve, we use a divide-and-conquer strategy: construct separate interpolations for each of $x(t)$, $y(t)$, and $z(t)$. From what we have learned about the parametric representation, this is all we need to construct a representation for the overall space curve.

We begin by looking only at the x-component of the fly's trajectory. This means that we shall be relating the x component of our trajectory data to the t component, and so the data to be considered can be reduced to

$$(t_0,x_0),\ (t_1,x_1),\ (t_2,x_2),\ \cdots,\ (t_n,x_n). \tag{6}$$

Because $t_{i+1} > t_i$, there will be no duplicate points, and we can actually plot the result on a standard Cartesian graph by plotting only (t,x) values.

The *Lagrange polynomial* $x(t)$ of degree $\leq n$ that passes through this data is defined as

$$x(t) = \sum_{i=0}^{n} x_i \prod_{j=0,\ j\neq i}^{n} \frac{t - t_j}{t_i - t_j}. \tag{7}$$

The result is therefore an explicit function x that varies with t. The product term (denoted by $\prod$) involving the t values accommodates unequally spaced samples in t. Later, when we talk about piecewise representations, the specific positions of the samples in t will be called *knots*. Sometimes the words *uniform* and *nonuniform* are used instead of *equal* and *unequal* with regard to knot spacing.

Example 2.1. If our sample data is:

$$(0,0),\ (1,1),\ (2,2),\ \cdots,\ (n,n), \tag{8}$$

that is, the fly travels in a straight line, then indeed $x(t) = t$. We therefore see that it is quite possible for $n+1$ samples to yield a "curve" of degree less than n.

Exercise 2.2. Verify the previous example by hand using the sample data up to $n = 2$. Even for such simple cases it is a somewhat tedious exercise, but it is worth doing.

Exercise 2.3. A much harder exercise is to prove inductively that for data of the form given in Eq. 8, the result is always a line of the form $y = x$.

Exercise 2.4. If the Lagrange polynomial must interpolate, among others, the point $(0,0)$, then coefficient a_0 of the resulting polynomial must be zero. Why?

Example 2.2. Suppose our data is $(-1,1), (0,0), (1,1)$. These points are conveniently chosen to lie on the curve $x = t^2$. Applying Lagrange interpolation,

$$p(t) = 1 \times \prod_{j=1,2} \frac{t - t_j}{-1 - t_j} + 0 \times \prod_{j=0,2} \frac{t - t_j}{0 - t_j} + 1 \times \prod_{j=0,1} \frac{t - t_j}{1 - t_j}$$

$$= \left(\frac{t}{-1} \cdot \frac{t-1}{-2}\right) + \left(\frac{t+1}{2} \cdot \frac{t}{1}\right)$$

$$= \frac{t^2 - t}{2} + \frac{t^2 + t}{2}$$

$$= t^2.$$

Exercise 2.5. The procedure **interp** in Maple performs Lagrange interpolation. Without looking at it, write your own interpolation procedure **lagrange(x,y,t)** where **x** and **y** are lists of values such that the points **[x[i],y[i]]** are to be interpolated, and **t** is the indeterminate of the resulting polynomial. If, as in the previous exercise, our data to be interpolated is $(-1,1)$, $(0,0)$, $(1,1)$, and we wish to compute $p(t)$, then a sample interaction with your procedure would be

```
> T := [-1,0,1];
                        T := [-1, 0, 1]
> X := [1,0,1];
                        X := [1, 0, 1]
```

```
> lagrange(T,X,t);
```
$$t^2$$

Exercise your code and **interp** on various data by displaying the resulting polynomials and by plotting them. Can you rewrite your procedure so that it avoids the use of loop constructs? Hint: use **prod** and **sum** on sequences.

Suppose the first six samples of our data are

$$(0,0),\ (1,2),\ (2,4),\ (3,1),\ (4,0),\ (5,-1).$$

Then the Lagrange polynomial that interpolates this data is

$$p_6(t) \ = \ -\frac{7}{40}t^5 + \frac{9}{4}t^4 - \frac{239}{24}t^3 + \frac{67}{4}t^2 - \frac{103}{15}t.$$

This function is plotted in Figure 2.3, and does not look bad. The desired sample points are plotted as small squares, and it is clear that the curve interpolates these values. Note that the interpolating polynomial predicts that in going from (0,0) to (1,2), the fly first goes downward and then moves upward. While this may actually be the case, and we have no evidence to the contrary, it is not the most direct trajectory. On the other hand, we do not expect flies to act rationally, so perhaps our model is satisfactory. The point here is that the data is not of itself sufficient to determine the some desirable qualities of the interpolation. The quality of the interpolation is partially a function of our sometimes subjective expectations. Apart from the fact that the coefficients are slightly disturbing in size, given that we have only provided six sample points, this interpolation appears smooth and convincing, disregarding that initial dip in the path.

Now suppose we would like to compute a partial trajectory on-the-fly (so to speak) before getting the entire list of sample points. This is an entirely legitimate desire, since our FlyTracker is a realtime device that feeds us a stream of data. It seems reasonable to attempt to compute an updated trajectory as we get new samples. After all, if our aim is to be able to continuously monitor the fly's position in the hope of swatting it, we will have failed in our task if we can't start monitoring until after all the data has been received and the fly has left the room. If we append the data point (6,1) to the list, we get the following polynomial:

$$p_7(t) \ = \ \frac{7}{144}t^6 - \frac{217}{240}t^5 + \frac{919}{144}t^4 - \frac{1003}{48}t^3 + \frac{2165}{72}t^2 - \frac{127}{10}t.$$

The coefficients are getting distressingly large. If we then add another point (7,0) to the list, for a total of eight samples, we get the seventh-degree polynomial

$$p_8(t) \ = \ -\frac{1}{80}t^7 + \frac{14}{45}t^6 - \frac{371}{120}t^5 - \frac{9887}{240}t^3 + \frac{18763}{360}t^2 - \frac{217}{10}t.$$

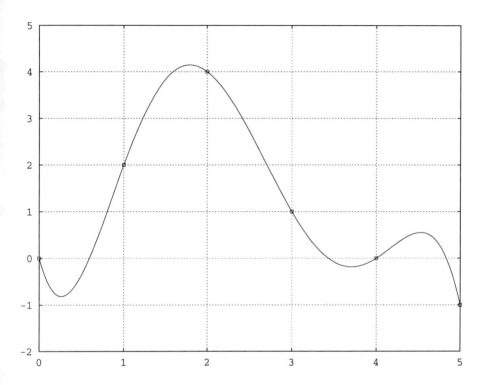

Figure 2.3.   Interpolating Lagrange polynomial $p_6(t)$ over six sample points.

Things are clearly getting out of hand. For one thing, a realtime tracking device can present data at a rate of 15 or more samples per second. After running our tracker for only ten seconds we would need a 149th-degree polynomial to plot the trajectory. Worse still, the size of the coefficients is growing at an alarming rate. Keeping an exact representation is out of the question. Even worse still is the news that comes with Figure 2.4, which plots p_6, p_7, and p_8 on the interval [0,5] for which they share the same samples. The trajectories only agree on the sample points and nowhere else! Moreover, notice that the dip in each polynomial for $t \in [0,1]$ gets deeper with increasing degree. The fly would thus travel a substantially different distance along each polynomial. The distance travelled can be measured by computing the arc length of each polynomial over any relevant interval for the parameter t. We shall discuss arc length in more detail in Chapter 4 on numerical integration; for an explicit differentiable function $x = f(t)$, the *arc length* of f on interval $[a,b]$ is defined as

$$a_f = \int_a^b \sqrt{1 + f'(t)^2}\, dt.$$

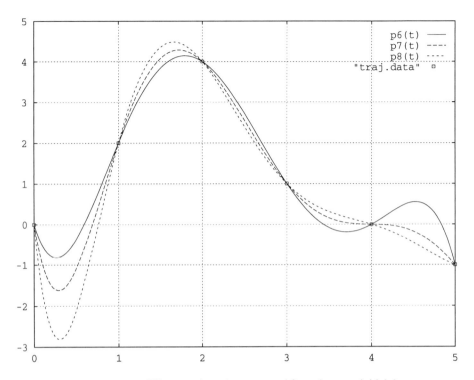

Figure 2.4. Three different trajectories computed from the same initial data.

Numerically computing this quantity for p_6, p_7, and p_8 on [0,5], we find that

$$a_{p_6} = 13.87, \quad a_{p_7} = 14.47, \quad a_{p_8} = 17.02,$$

which confirms our visual suspicions. Clearly Lagrange interpolation cannot be used in the successive computation of partial trajectories. On the other hand, it is quite useful for solving some integration problems, as we shall see in Chapter 4.

Example 2.3. When Maple executes the following script (or use your **lagrange** code):

```
T := 0,1,2,3,4,5:
X := 0,2,4,1,0,-1:

p6 := interp([T],[X],t);
```

we get

$$p6 := -\,7/40\ t^5\ + 9/4\ t^4\ - 239/24\ t^3\ + 67/4\ t^2\ - 103/15\ t$$

which corresponds to $p_6(t)$ above.

Example 2.4. To compute the arc length of a polynomial $p(t)$ on an interval $[a,b]$ in Maple, we can try:

```
int ( sqrt(1 + diff(p,t)^2), t=a..b );
```

However, a closed-form solution for this integral rarely exists. What is the highest degree for which Maple can find a closed-form solution? (Hint: it is a small number.) We can compute the result numerically by stating

```
evalf( Int ( sqrt(1 + diff(p,t)^2), t=a..b ) );
```

To get the arc length of p_6:

```
evalf(Int( sqrt(diff(p6,t)^2+1), t=0..5));
```

$$13.87427367$$

Later, we shall discuss problems that are common to all interpolation schemes: ringing and instability in the face of "noisy" data. The following two exercises are an instructive introduction.

Exercise 2.6. Write a Maple program to generate a sets of "noisy" points about the line $x = t$. In particular, generate points of the form $(t, t+\varepsilon)$, where $t \in \mathbf{N}$ and ε is a small randomly chosen displacement in the range $[-0.1, 0.1]$. Compute the interpolating polynomial for this data, and plot it against the line segment $x = t$ in the relevant range. Another useful experiment is to put all points on the line $x = t$ (or any other line) and to perturb the x value of one of the interior points. What is the degree of the polynomial passing through the perturbed data?

Exercise 2.7. Write a Maple program to generate a set of data of the following form:

$$(0,0), (1,0), (2,0), \cdots, (k, 0), (k+1,1), (k+2,1), (k+3,1), \cdots, (2k, 1),$$

for a given $k \in \mathbf{N}$. Such a set of samples could be derived, for example, from an electrical signal carrying, at any point in time, either 0 or 1 volts. Plot the interpolating Lagrange polynomial for various values of k. What happens to the polynomial around $(k, 0)$?

Lagrange interpolation is remarkable in that it prescribes a way of generating a single polynomial to pass through an arbitrary set of sample points. However the result may be unwieldy. This is not to say that Lagrange interpolation is hopelessly flawed for our purposes. The specific conclusions we can make are:

- Deriving a single interpolating polynomial for a large set of samples is not feasible especially if the interpolation is to be performed *online*.
- An interpolation must be predictable and stable. If we make a small change or add a sample point we expect a small change in the interpolated trajectory. Lagrange interpolation does not have this property.
- If we wish to use polynomial interpolants, then we should search for lower-degree approaches.

- For many applications, the ability to perform interpolation before all of the sample data are known is very important, so we require a scheme that does not affect the "past."

There is no need to panic. There exist interpolating polynomial solutions that satisfy all of these problems, but we must abandon the idea of a single global solution for our trajectory problem and instead look for *local* solutions that can be pieced together to form an overall solution. This is yet another instance of a divide-and-conquer strategy in a slightly disguised form.

Exercise 2.8. Now that you know how to compute a polynomial $x(t)$, outline an algorithm to generate a parametric space curve $(x(t), y(t), z(t))$ using Lagrange interpolation.

Example 2.5. Recall that the formula for Lagrange interpolation does not assume equally spaced points in the parameter. The reader is encouraged to use Maple to verify the following simple derivation. Let $P_0(u_0, x_0)$ and $P_1(u_1, x_1)$ be two points in $\mathbf{R}^2$ with $u_1 > u_0$. Use Maple or your ingenuity to show that the (linear) Lagrange polynomial that interpolates x_0 and x_1 is

$$x(u) = \frac{x_0 - x_1}{u_0 - u_1} u + \frac{x_1 u_0 - x_0 u_1}{u_0 - u_1}.$$

To get the line segment between x_0 and x_1, we must restrict u to be in $[u_0, u_1]$. Suppose we now wish the resulting function to be parameterised on $[0,1]$. The last chapter showed us how to do this: introduce a new parameter

$$t = \frac{u - u_0}{u_1 - u_0},$$

meaning that

$$u = t(u_1 - u_0) + u_0.$$

(Notice that this expression itself defines a linear interpolation between u_0 and u_1 for $t \in [0,1]$.) We substitute this expression for u into $x(u)$, and after simplifying the rather messy subexpressions, we arrive at

$$x(t) = x_0 + t(x_1 - x_0)$$

with $t \in [0,1]$. Thus the definition we have been using all along for line segments parameterised on $[0,1]$ is exactly consistent with the result of Lagrange interpolation. This result should help to convince the reader that we are not unduly restricting ourselves if we use a consistent parameterisation over $[0,1]$.

2.4. Piecewise Polynomial Interpolation

One way to accommodate creating an overall solution from local solutions is to construct a set of polynomial *curve segments* that each interpolate a subset of the sample data. If we are careful, we might be able to stitch these solutions together into a reasonable whole. Putting curve segments together into a global solution is called, naturally enough, *piecewise polynomial interpolation.*

In what follows, we shall sacrifice mathematical rigour for intuition. We shall first illustrate the basic concepts of piecewise polynomial curves by graphical examples. We then discuss some of the many possible polynomial representations for individual curve segments. Finally, we show how to piece the curve segments together. Some of the basic mathematics will then be presented, but we will of necessity be leaving large gaps.

One issue of interest is the choice of the degree of the interpolating polynomials. Recall our remark that very low-degree polynomials may not be flexible enough, but very high-degree polynomials wiggle too much. The most common compromises are to use second-degree polynomials (the *quadratics*) or third-degree polynomials (the *cubics*). The extra degree in cubics over quadratics allow one to define space curves that are non-planar within a single curve segment. Cubics can also be connected together more smoothly. Depending on the application, this may be useful. The price paid is slightly more expensive evaluation. An application may demand higher- or lower-degree forms, and indeed a mixture of curve segments of various degrees is possible. Let us first see how low-degree polynomials can be applied to our trajectory problem.

2.4.1. Piecewise Linear Interpolation

If, as before, we have a simple sequence of samples

$$(t_0,x_0),\ (t_1,x_1),\ (t_2,x_2),\ \cdots,\ (t_n,x_n), \tag{9}$$

then one interpolation strategy would be to draw line segments between consecutive points. More precisely, the overall "curve" would consist of line segments of the following form:

$$L((t_i,x_i),\ (t_{i+1},x_{i+1})),\ \ i = 0, 1,\ \cdots, n-1. \tag{10}$$

We write "L" rather than "$\mathbf{L}$" because we are dealing with only with the x component of a curve. This notation in the previous chapter denoted the set of points on the line segment. As we saw above, the Lagrange polynomial between two or more points lying on a line is in fact that line, so this form of interpolation can be called *piecewise linear Lagrange interpolation.* A more economical name is *piecewise linear interpolation.* In our work we shall be using line segments rather than lines, but this amounts to using a line over a finite parametric interval.

With the familiar sequence of samples

$$(0,0),\ (1,2),\ (2,4),\ (3,1),\ (4,0),\ (5,-1), \tag{11}$$

we would expect to get the trajectory in Figure 2.5. Notice that the samples are equally spaced in t, but we now know that this is unnecessary. This is obviously not a natural trajectory for a fly to follow, because presumably it is hard for a fly to turn corners so abruptly, nor, putting ourselves in the fly's position, would it be

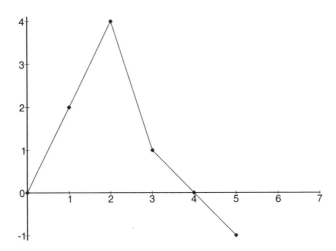

Figure 2.5. Piecewise linear interpolation on six samples.

easy to fly in such straight lines. However, this approach has the nice advantage that the past trajectory does not change if we add more sample data. In particular, after adding (6,1) followed by (7,0) to the list, we get the trajectories in Figures 2.6 and 2.7, leaving the trajectory over the first five samples unchanged.

We do not yet have a realistic model for fly trajectories, but we shall soon remedy this. It is worth making a few remarks. First, if sample points are spaced very closely, then a linear interpolation may be visually satisfactory. Second, if our aim were not to compute a trajectory but instead to approximate the integral of a function f passing through a set of points, then an approximation of f by linear interpolation and its subsequent integration is in fact the *compound trapezoid rule*, as will be seen in Chapter 4. In this case, sharp corners in the function are undesirable only if they cause inaccuracy in the integral. Once again, we see that the desirable qualities of an interpolation depend on its use.

If a piecewise linear fly trajectory is unsatisfactory, we could try interpolating using higher-order polynomials. A good choice is the cubic polynomials. Some visual evidence for this choice can be seen in Figure 2.8. The piecewise cubic trajectory appears smooth and its corners are rounded. If we add the same extra points, we see in Figure 2.9 that this interpolation changes the past only slightly at the join between curve segments. Figure 2.10 compares piecewise linear and cubic interpolations with global Lagrange interpolation. Notice the larger turns made by Lagrange polynomial, and that the piecewise cubic curve stays close to the linear trajectory. The arc length of the cubic curve is 14.42, while the ''arc length'' of the straight trajectory (i.e., the sum of the lengths of the line segments) is 14.11. Recall that the Lagrange polynomial p_8 has arc length 17.02.

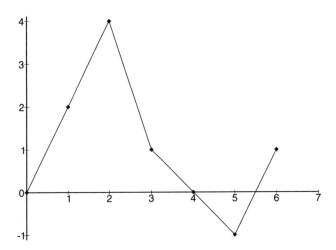

Figure 2.6. Piecewise linear interpolation on seven samples.

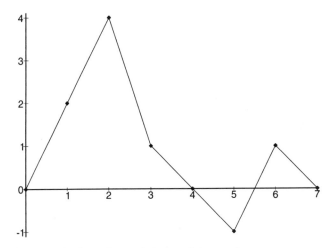

Figure 2.7. Piecewise linear interpolation on eight samples.

2.4.2. *Representations for Polynomial Curves*

There are many ways of expressing polynomials. Why do we need more than one way? Because each way gives us another point of view. By way of analogy, let us return to Figure 2.1. There we see that a curve has been defined in the co-ordinate system of a tracking device that has a very specific notion of *up, right,*

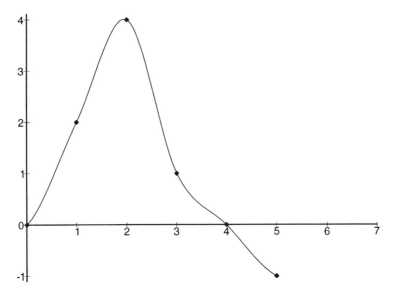

Figure 2.8.   Interpolating piecewise cubic trajectory.

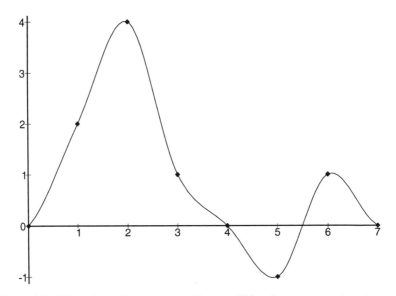

Figure 2.9.   Piecewise cubic trajectory with two additional curve segments.

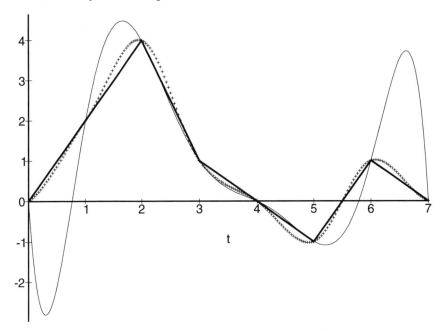

Figure 2.10. A comparison of three interpolating trajectories: degree-seven Lagrange (thin line), piecewise cubic (hashed line), and piecewise linear (thick line).

out, and an origin. But this does not constrain us from *viewing* the data in different ways by imposing our own co-ordinate system on top of this data. This effectively causes us to relabel all the data from the tracker. If, for example, the tracker were turned on its side, then we would have to relabel the co-ordinate axes differently to properly reinterpret our idea of *up*, *right*, and *out*. We may also need to scale the data or perform other transformations on it. The moral is that there are many ways to represent the same position in space, but that they (usually) differ from one another by a simple transformation.

Exactly the same is true of the polynomials. We have so far defined only one representation, but we shall see that there are many ways of writing the same polynomials, and each representation differs from another only by a simple transformation. But we still haven't properly answered our original question. While multiple representations seem to be reasonable for points in space, why should the argument carry over to polynomials? The answer is that different representations give us different "handles" on the curves and surfaces we are manipulating: some representations are particularly good for shaping the object, others for controlling its smoothness, and yet others are good for efficiently evaluating points on the object. We can exploit different representations for their different features, transforming appropriately as we go along.

A little more formally, when we write down the expression $P(x,y,z) \in \mathbf{R}^3$, we mean that P is some point in space. One way to think of a point is to consider its corresponding *position vector* $\mathbf{P}$, which is the vector extending from the origin to the point P. Mathematically, the position vector $\mathbf{P}$ is defined as

$$\mathbf{P} = x\,\mathbf{e}_x + y\,\mathbf{e}_y + z\,\mathbf{e}_z \tag{12}$$

$$= [\mathbf{e}_x \ \mathbf{e}_y \ \mathbf{e}_z] \begin{bmatrix} x \\ y \\ z \end{bmatrix}, \tag{13}$$

where $(\mathbf{e}_x, \mathbf{e}_y, \mathbf{e}_z)$ are the *basis vectors* for our familiar Cartesian space, namely

$$(\mathbf{e}_x, \mathbf{e}_y, \mathbf{e}_z) = ((1,0,0), (0,1,0), (0,0,1)). \tag{14}$$

These basis vectors represent the directions (*right, up, out*) of our co-ordinate axes and are of unit length. We shall think of a point and its position vector as denoting the same object, (see Figure 2.11 left).

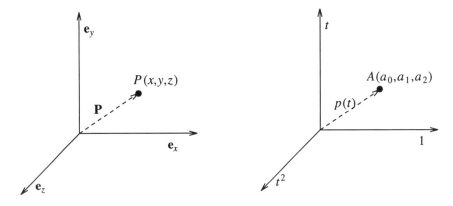

Figure 2.11. Left: A point $P(x,y,z)$ and its position vector $\mathbf{P}$. Right: A polynomial $p(t)$ and by its "position" $A(a_0,a_1,a_2)$.

Polynomials can be expressed in exactly the same way as points in space. From Example 1.6 of the previous chapter, we can write down a polynomial of degree n in matrix-vector form. For example, a polynomial of degree n over parameter t, which we have written as

$$p(t) = a_0 + a_1 t + a_2 t^2 + \cdots + a_n t^n \tag{15}$$

can be written in vector form as

$$p(t) = [t^n \ t^{n-1} \ \cdots \ t^1 \ t^0] \begin{bmatrix} a_n \\ a_{n-1} \\ \vdots \\ a_0 \end{bmatrix}. \tag{16}$$

For short, we shall write this as

$$p(t) = TA, \qquad (17)$$

where T is a row vector of *basis polynomials* and A is a column vector of coefficients. There is nothing special about the particular form of the vector (and later matrix) multiplication and the specific way in which we have written the elements of T and A. There is also nothing special about the choice of polynomials as a basis: other kinds of functions can also serve as bases. We shall see, for example, that Fourier series have *basis functions* consisting of sines and cosines.

There is a useful way of looking at Eq. 17: just as the point P above is a point with co-ordinates (x, y, z) in a Euclidean space that happens to be very familiar (see Eq. 13), so too $p(t)$ is a point A in a *polynomial function space*. The set of polynomials of degree n lie in an $n+1$-dimensional polynomial function space. This is to say that the dimension coincides with the order of the polynomial. Figure 2.11 (right) gives a representation for a quadratic polynomial in a familiar polynomial function space. The axes are drawn as being orthogonal (just as in the spatial case at left), but in reality these basis polynomials are not mutually perpendicular.[1] As the degree of the polynomial increases, it becomes difficult to visualise it as a point, but that is no different from trying to graph any other kind of point in a multidimensional space.

Exercise 2.9. If a cubic polynomial can be represented as a four dimensional point in a polynomial function space, then how would you represent a polynomial space curve $(x(t), y(t), z(t))$ in which each of x, y, z are cubic polynomials?

We will not deal at length with the mathematics of polynomial function spaces, but it is very important to get a feeling for this representation. It means that just as we can change our point of view of the spatial point P and get different co-ordinates for P, we can change our representation of polynomial $p(t)$. We shall see several examples of this shortly. Furthermore, we can perform transformations on the same representation. For example, suppose P is the position of our fly at rest on the wall, where the co-ordinates (x, y, z) for P are measured from the point of view of our tracker. If we were to pick up the tracker and move it across the room, or turn it upside-down, then the co-ordinate system for the tracker would change with respect to the room, and the value reported for the position of the fly will change. But from our point of view, the fly is actually at the same position in space, resting on the wall. Because we know the orientation of the tracker, we can consistently maintain this representation. Similarly, the same polynomial can be written down in different ways using

---

1. The notion of orthogonality can be generalised to functions by extending the dot product operation for vectors to an integral. We shall see examples of orthogonal functions in Chapter 5.

different polynomial bases. Equations 4 and 16 reflect the representation of polynomials using the standard *monomial* or *power* basis. The basis polynomials for the monomial form are given by T in Eq. 17 for the desired degree n.

We can write down (for what must be the nth time) the equation for a line segment between points $P_0(x_0, y_0, z_0)$ and $P_1(x_1, y_1, z_1)$ as follows:

$$\mathbf{L}_1(t) \;=\; (1{-}t)P_0 + tP_1 \,, \quad t \in [0,1] \tag{18}$$

$$= \; [(1{-}t) \;\; t] \begin{bmatrix} P_0 \\ P_1 \end{bmatrix}$$

$$= \; [(1{-}t) \;\; t] \begin{bmatrix} x_0 & y_0 & z_0 \\ x_1 & y_1 & z_1 \end{bmatrix}.$$

We have made explicit the fact that in $\mathbf{R}^3$, this is a specification for three separate functions. That is $\mathbf{L}_1(t) = (x(t), y(t), z(t))$, where

$$x(t) \;=\; [(1{-}t) \;\; t] \begin{bmatrix} x_0 \\ x_1 \end{bmatrix},$$

$$y(t) \;=\; [(1{-}t) \;\; t] \begin{bmatrix} y_0 \\ y_1 \end{bmatrix},$$

$$z(t) \;=\; [(1{-}t) \;\; t] \begin{bmatrix} z_0 \\ z_1 \end{bmatrix}.$$

The representation of the same line segment in the power basis is

$$\mathbf{L}_2(t) \;=\; P_0 + t(P_1 - P_0) \tag{19}$$

$$= \; [t \;\; 1] \begin{bmatrix} P_1 - P_0 \\ P_0 \end{bmatrix}.$$

It is helpful to look at graphs of the functions found in the vector T; we have called these functions *basis* functions because they act in a manner analogous to basis vectors in a Cartesian space. For good reason, these functions are also called *blending* functions. The overall curve for any value of the parameter(s) is defined by multiplying the value of the blending function with the relevant data value, and summing the result. In fact, a polynomial $p(t)$ of degree n can be represented as

$$p(t) \;=\; \sum_{i=0}^{n} \alpha_i \, b_i(t), \tag{20}$$

where the $b_i(t)$, $i = 0, 1, \cdots, n$ are the $n+1$ basis functions, and the α_i are the corresponding data items.

Example 2.6. In the power-basis representation, $b_i(t) = t^i$, and $\alpha_i = a_i$. Why do you think this is also called the *monomial* form?

Exercise 2.10. Recalling the definition of Lagrange interpolation in Eq. 7, what are the relevant α_i and b_i? Assume the spacing of knots in t is uniform, say one unit. Use Maple to derive the Lagrange basis functions for polynomials of degrees two to six. These are called the *Lagrange polynomials* of degree n. Plot each set of basis functions on an appropriate domain.

Notice that the data items are constant for each polynomial. For a line segment, P_0 and P_1 are constant over that segment. Only the values of the blending functions change. In representation $\mathbf{L}_1$, the blending function $1-t$ controls the contribution of P_0 to the result for any value of $t \in [0,1]$, and similarly for effect of t on P_1. Figure 2.12 depicts the blending functions $[1-t \quad t]$ and $[1 \quad t]$.

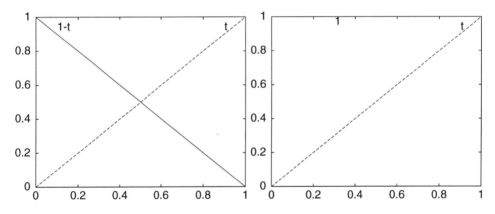

Figure 2.12. Blending functions: at left $[1-t \quad t]$; at right $[1 \quad t]$.

We hinted above that different representations have different uses. For line segments, the issue is not crucial, though even in this case recall from our discussion in the last chapter that there are good reasons for looking at line segments in different ways. In particular, we used the power-basis form of Eq. 19 (and Eq. 20) to derive a line-rendering algorithm. We shall similarly see that to render higher-order polynomial curves, the power basis is a useful normal form. The case for multiple representations becomes much clearer for higher-order polynomials. It is hard, for example, to manipulate a curve directly from its coefficients. We shall illustrate this fact by the following exercise.

Exercise 2.11. Let us give the first four sample points of our fly's trajectory to Maple as follows.

```
> x1 := [0,1,2,3];
```
$$x1 := [0, 1, 2, 3]$$

```
> y1 := [0,2,4,1];
```
$$y1 := [0, 2, 4, 1]$$

```
> pt := interp(x1,y1,t);
```
$$pt := - 5/6\ t^3 + 5/2\ t^2 + 1/3\ t$$

```
> plot(pt,0..4);
```

Thus we get the polynomial

$$p(t) \;=\; -\frac{5}{6}t^3 + \frac{5}{2}t^2 + \frac{1}{3}t$$

and the plot in Figure 2.13. Using this polynomial as a starting point, suppose we now would like to change the curve so that its left half is less straight and arches more like, say, a cosine function. Hard: show how to do this by directly manipulating the coefficients of $p(t)$. There is at least one much easier way. What is one way?

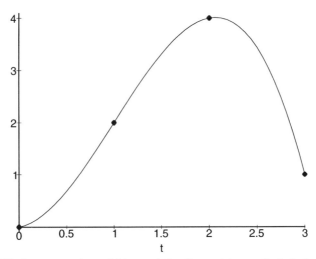

Figure 2.13. Lagrange polynomial interpolating four points on a fly trajectory.

This exercise demonstrates that there is little obvious connection between a polynomial's coefficients and the finer aspects of its shape. Another approach would be to change the representation of the curve into one that is better suited to a particular class of such changes, make the changes, and if necessary transform back to the original representation. In some cases, alternative basis functions can be generated directly. For example, in Exercise 2.10, we saw that the formula for Lagrange interpolation allows us to compute directly the Lagrange basis polynomials of any desired degree.

Generating basis functions directly is not always easy. An alternative strategy is to perform a *change of basis*: derive a transformation that rewrites a polynomial using different basis functions. Mathematically, a change of basis is easy to describe. We simply want another way to represent our $n+1$ coefficients A in terms of a comparable set of $n+1$ independent "observations" V. It turns out that a single matrix can reflect this representation as follows:

$$A = M \begin{bmatrix} V_0 \\ V_1 \\ \vdots \\ V_n \end{bmatrix} \tag{21}$$

$$= M V,$$

where M is a specific $(n+1) \times (n+1)$ matrix for polynomials of degree n. Putting this new formula for A into Eqs. 16 and 17, we get

$$p(t) = T M V. \tag{22}$$

If we multiply $T M$ together, we get a different set of polynomial basis (or blending) functions. Thus, the set of observations V is really a *point* in the new polynomial function space for our representation. The reader may prefer to view this as an algorithm for specifying a desired behaviour using V and mapping to the monomial form using M.

Example 2.7. At times, it might be preferable to view a line segment as in representation L_1 (Eq. 18). We can relate the two line-segment representations discussed above as follows:

$$L(t) = T M V$$

$$= [t \quad 1] \begin{bmatrix} -1 & 1 \\ 1 & 0 \end{bmatrix} \begin{bmatrix} P_0 \\ P_1 \end{bmatrix}.$$

The matrix gives the change of basis from the form of L_1 to L_2. We can see this if we simply multiply $M V$:

$$M V = \begin{bmatrix} P_1 - P_0 \\ P_0 \end{bmatrix}.$$

On the other hand, the polynomial blending functions in representation L_1 can be seen by multiplying $T M$:

$$T M \ = \ [(1-t) \ \ t].$$

At this point the reader may wonder why such an indirect route is being used. If, after all, what we really want is different basis functions, why not just define them directly as we could with Lagrange interpolation? The reason is that defining basis functions to achieve a particular effect is usually no easier than toying with coefficients of a polynomial to get a desired shape. It is far easier to *specify* the desired behaviour of a curve and to *deduce* the basis functions by means of the change of basis. We now show how this is done.

For higher-degree forms, there are many different and useful polynomial representations, but we shall focus only on the interpolating ones that might be useful to our problem of modelling trajectories. In that case, we need a cubic polynomial that will interpolate points in space. It is quite possible to define a form that interpolates four values. For example, if we wish a polynomial $p(t)$ defined over $t \in [0,1]$ to interpolate four values x_0, x_1, x_2, x_3 subject to

$$p(0) = x_0, \quad p\left(\frac{1}{3}\right) = x_1, \quad p\left(\frac{2}{3}\right) = x_2, \quad p(1) = x_3, \tag{23}$$

then by solving a system of equations,[2] we find that the appropriate change of basis matrix is

$$M_I \ = \ \begin{bmatrix} -\dfrac{9}{2} & \dfrac{27}{2} & -\dfrac{27}{2} & \dfrac{9}{2} \\[2mm] 9 & -\dfrac{45}{2} & 18 & -\dfrac{9}{2} \\[2mm] -\dfrac{11}{2} & 9 & -\dfrac{9}{2} & 1 \\[2mm] 1 & 0 & 0 & 0 \end{bmatrix}. \tag{24}$$

This means that we can represent a four-point cubic interpolant as

$$p(t) \ = \ [t^3 \ \ t^2 \ \ t \ \ 1] \, M_I \begin{bmatrix} x_0 \\ x_1 \\ x_2 \\ x_3 \end{bmatrix}. \tag{25}$$

---

2. The technique for solving this system is outlined in Appendix A.

The blending functions for this basis are given by $T M_I$. Figure 2.14 depicts these functions. Notice that at four distinct points for t, namely $t = 0$, $1/3$, $2/3$, and 1, one of the blending functions has the value one while the others are zero. When the blending functions for an interpolating polynomial scheme have this property, the functions are called called a *cardinal basis*. As it happens, this formulation of the four-point interpolation is exactly the Lagrange cubic interpolant on four equally spaced points over $[0,1]$!

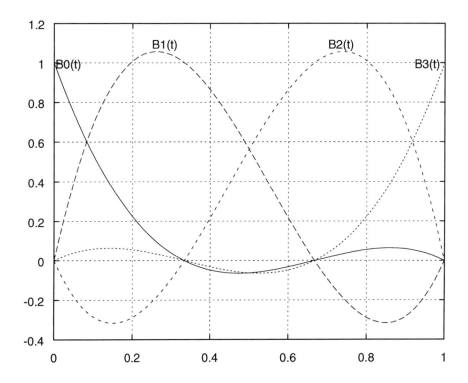

Figure 2.14. Blending functions for the cubic four-point interpolant.

Exercise 2.12. Verify that this form does interpolate the points P_i, $i = 0,1,2,3$.

Exercise 2.13. Revisit Exercise 2.10 to show that the polynomial $p(t)$ resulting from the four-point interpolant M_I coincides with the cubic interpolating Lagrange polynomial. Should we be surprised by this coincidence?

The difficulty with the above form is that it leaves us no room to manoeuvre. If our aim were only to interpolate four points, then we would be in fine shape. However, our final goal is to be able to interpolate a continuing sequence of values and to stitch together a smooth trajectory in terms of curve segments. A cubic solution that interpolates four points will force the stitching to have sharp corners on it, as we will discuss shortly.

For our purposes, a good solution would be one that both interpolates and admits easy piecing together of curve segments. There exist many interpolating formulations with this property. In this regard, the cubic *Overhauser* or *Catmull-Rom* basis is a very good one. It has the following basis matrix M_{CR}:

$$M_{CR} = \frac{1}{2} \begin{bmatrix} -1 & 3 & -3 & 1 \\ 2 & -5 & 4 & -1 \\ -1 & 0 & 1 & 0 \\ 0 & 2 & 0 & 0 \end{bmatrix}. \tag{26}$$

This basis has the blending functions depicted in Figure 2.15. Let us return to our sample data above. The first four points of the data were

$$(0,0),\ (1,2),\ (2,4),\ (3,1). \tag{27}$$

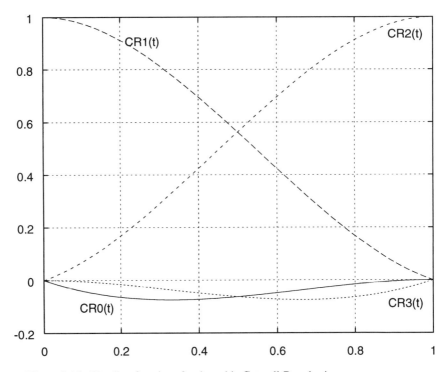

Figure 2.15. Blending functions for the cubic Catmull-Rom basis.

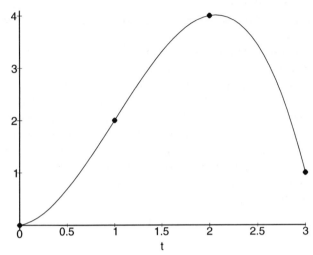

Figure 2.16. The result of applying the four-point interpolating cubic, M_I.

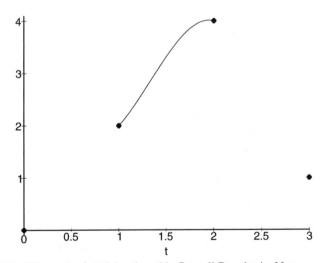

Figure 2.17. The result of applying the cubic Catmull-Rom basis, M_{CR}.

Applying the four-point interpolating cubic basis given by M_I to these samples gives the expected result in Figure 2.16. More surprising is the effect of applying the Catmull-Rom basis M_{CR} to the same samples. As seen in Figure 2.17, only the two central points are interpolated. Contrasting the basis functions for four-point interpolation in Figure 2.14 with those for cubic Catmull-Rom interpolation in

Figure 2.15 gives an immediate explanation: unlike the four-point basis functions, among the Catmull-Rom basis functions there are only two points in which three curves are zero and the other is one, namely at $t = 0$ and $t = 1$. Interpolation can only occur at these values of the parameters. Thus the Catmull-Rom basis is in cardinal form. Moreover, the range of the interpolation is clear: the middle two basis functions are those which achieve the value one while the others are zero. Thus interpolation can only occur between the middle two points.

Why is this behaviour in the least desirable, when the other basis interpolates all four points? The answer is that by sacrificing a little in interpolation power on each curve segment, we gain greater flexibility when we stitch curve segments together to get a final trajectory. Figure 2.18 depicts the result of two curve segments after having added one extra sample, (4,0). The first curve segment is still defined over the points in Eq. 27, while the second curve segment is defined over a "right shift" of these points:

$$(1,2), (2,4), (3,1), (4,0). \tag{28}$$

Thus for any curve segment p_i, only four sample points in the trajectory are used, and moreover, the first three of those sample points are also used to define p_{i-1} and the last three are also used for p_{i+1}. This ensures that the "seams" between the curve segments are invisible and that no cusps are introduced in the approximation, assuming all interpolated points P_j are distinct. This is no accident: the Catmull-Rom basis is specifically designed to have this behaviour. A large class of polynomial bases called the *B-splines* have this behaviour.

In contrast to the Catmull-Rom form, we can apply M_I to the two sets of sample points (0,0), (1,2), (2,4), (3,1), and (1,2), (2,4), (3,1), (4,0). As might be expected, we should get two distinct curve segments, not a seamless trajectory. This behaviour is illustrated in Figure 2.19. A more appropriate way to use this basis is to interpolate sets of four samples such that only the endpoints are shared by adjacent curve segments. That is, if we have points $a, b, c, d, e, f, g, h, i, k$ to interpolate, we can create three curve segments defined, respectively, over the following sets of four samples:

$$(a, b, c, d), (d, e, f, g), (g, h, i, k).$$

Figure 2.20 depicts a piecewise cubic curve trajectory rendered in this fashion. Note the cusp that appears at the joint between the first and second curve segments (at point (3,1)). The overall curve is continuous everywhere, including the joint between segments, but the derivative at the joint is undefined. This problem is inherent to the four-point interpolant, because in defining it we imposed four independent constraints on the basis, namely that each of the four specified points must be interpolated. In a cubic basis, we have no other degrees of freedom available. On the other hand, the Catmull-Rom basis is defined so that two constraints are used in interpolating the middle two sample points, and two constraints define the derivative at the ends of the curve segment. The derivative

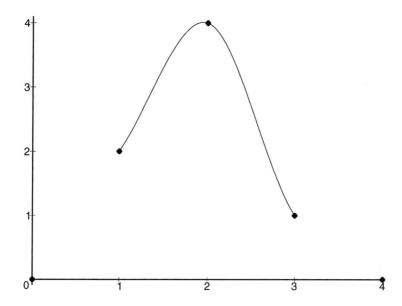

Figure 2.18. Two cubic Catmull-Rom curve segments stitched together.

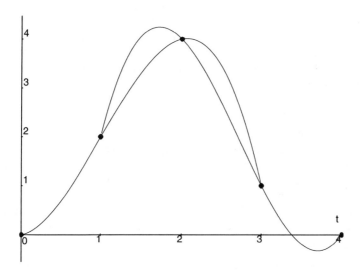

Figure 2.19. Two cubic four-point interpolant curve segments ''stitched'' together.

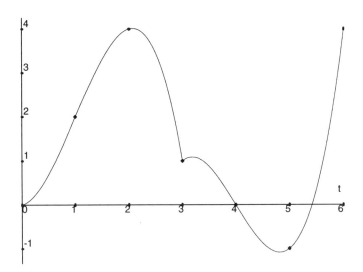

Figure 2.20. A better stitching job between two interpolating curve segments.

is arranged to be continuous across curve segments.

Note that there is nothing preventing us from treating each of the parametric functions $x(t), y(t), z(t)$ differently. We could, for example, employ Catmull-Rom interpolation for function x, the four-point interpolant for y, and linear interpolation for z. However, generally we would not like to tie the behaviour of a function to the co-ordinate system in which it is defined, but there are occasionally good reasons for doing this.

Exercise 2.14. Give some situations in which it may be preferable to use different curve specifications for the parametric functions.

Exercise 2.15. If we have n sample points to interpolate, $n > 4$, how many cubic Catmull-Rom curve segments are needed to perform the interpolation? How many four-point cubic interpolating curve segments are needed in the various ways they can be used? How many linear interpolating segments are needed?

Exercise 2.16. In Figure 2.18 (and others), the first and last sample points are not interpolated. However, in Figure 2.8 the first sample point *is* interpolated (as is the last). How is this accomplished?

Exercise 2.17. The horizontal axis of each plot involving Catmull-Rom interpolation is labelled "t" and varies from 0 to 3 or 4. But each Catmull-Rom polynomial is parameterised on $t \in [0, 1]$. Explain the apparent contradiction.

Exercise 2.18. Given five points $x_0, x_1, \cdots, x_4$, let $p_0(t)$ be the polynomial got by applying the Catmull-Rom basis to x_0, x_1, x_2, x_3 and likewise for $p_1(t)$ on x_1, x_2, x_3, x_4. Prove that $p_0(1) = p_1(0)$. For their derivatives p'_0 and p'_1, show also that $p'_0(1) = p'_1(0)$.

2.4.3. Putting the Pieces Together

In a global interpolation scheme such as Lagrange interpolation, a set of samples is taken as input, and a single polynomial p is the outcome. On the other hand, in a piecewise polynomial solution, we instead produce a sequence of *polynomial curve segments* $p_0, p_1, \cdots, p_m$. Each segment is defined over some small subset of the sample points. To construct a smooth trajectory, we must appropriately order the curve segments so that each curve smoothly connects to its immediate neighbours in the sequence. At any point in our solution, exactly one p_i is active, and thus we must also define the domain over which each curve segment is active. Our investment in parameterisation over $[0,1]$ will pay off. Let us first consider piecewise linear interpolation in the x component of a space curve.

Recall from Eq. 10 that if we have the x component of a set of samples

$$(t_0, x_0), (t_1, x_1), (t_2, x_2), \cdots, (t_n, x_n), \tag{29}$$

then a piecewise linear interpolation was given by a sequence of n (linear) curve segments $p_0(t), p_1(t), \cdots, p_{n-1}(t)$, such that when t varies from 0 to 1, each $p_i(t)$ will "draw" the line segment from (t_i, x_i) to (t_{i+1}, x_{i+1}). We observe that both the active curve segment p_i and the current value of t can be encoded as a single real number $r = i + t$. The whole part of r gives the current curve segment and the fractional part of r is the current value of t. Because $0 \le i \le n-1$, r is nicely in the real-valued range $[0, n]$. (This may be a good time to revisit Exercise 2.17.)

By following Example 2.7, we can write each $p_i(t)$ in matrix form

$$p_i(t) = T M X_i, \tag{30}$$

where $T = [t \ 1]$ is the power basis of degree one with $t \in [0,1]$, M is one of the linear interpolation bases (see above), and

$$X_i = \begin{bmatrix} x_i \\ x_{i+1} \end{bmatrix}.$$

For segment $i = 0$, for example,

$$p_0(t) = T M \begin{bmatrix} x_0 \\ x_1 \end{bmatrix}. \tag{31}$$

The pseudocode in Figure 2.21 implements linear interpolation in a single variable. We have deliberately expressed it in a form that will allow us to generalise to higher-order polynomials, and to allow variation in y and z as well as x. An efficient implementation of **Render** for line segments was discussed in the last chapter. Notice that if we drop the **Render** operation and store the p_i, then we can use them for other things, such as to compute integrals.

The approach for higher-order polynomial interpolation is identical: we just slide in new values for T, M, and X_i. An algorithm to derive the piecewise cubic Catmull-Rom interpolants is illustrated in Figure 2.22.

```
for i=0..n−1
```
$$p_i \; := \; T \; M \begin{bmatrix} x_i \\ x_{i+1} \end{bmatrix}$$
Render(p_i **for** $t \in [0,1]$)
```
end for
```

Figure 2.21. A naive rendering algorithm for piecewise linear interpolation.

```
for i=0..n−3
```
$\quad p_i \; := \; T \; M \; X_i$
$\quad$ **Render**(p_i **for** $t \in [0,1]$)
```
end for
```

$$\textbf{where} \; X_i \; := \begin{bmatrix} x_i \\ x_{i+1} \\ x_{i+2} \\ x_{i+3} \end{bmatrix}$$

Figure 2.22. Rendering piecewise cubic Catmull-Rom curves.

Exercise 2.19. Modify the Catmull-Rom interpolation algorithm to interpolate points using the four-point basis instead.

2.4.4. *General Space Curves*

So far, we have only looked at the x component of piecewise polynomial curves. The parametric approach will again be of great help, because it allows us to separate the complicated behaviour of the fly in $\mathbf{R}^3$ into (hopefully simpler) behaviours in each of x, y, and z. This allows us to almost trivially extend our approach to space curves.

Let us go back to our original assumption that the data values returned by our FlyTracker are of the form

$$(t_0, x_0, y_0, z_0), \; (t_1, x_1, y_1, z_1), \; (t_2, x_2, y_2, z_2), \; \cdots. \tag{32}$$

We shall assume that the samples come at a regular time spacing. That is, $t_i = i\Delta t$. Our formulations for four-point and Catmull-Rom interpolation assume uniform spacing, but there are many other (usually non-matrix) forms that allow nonuniform spacing. Modelling a general space-curve trajectory requires that we change our point of view slightly to the *parametric representation* of the points. Thus, rather than looking at the points as the quadruples in Eq. 32, we instead break them up into three separate sequences of ordered pairs:

$$(t_0,x_0),\ (t_1,x_1),\ (t_2,x_2),\ \cdots,$$
$$(t_0,y_0),\ (t_1,y_1),\ (t_2,y_2),\ \cdots, \tag{33}$$
$$(t_0,z_0),\ (t_1,z_1),\ (t_2,z_2),\ \cdots.$$

The time value is thus common to each sequence of ordered pairs.

To make the results easy to plot, we shall drop (without loss of generality) the z component from our solution. Suppose our sequence of samples (t,x,y) is

$$(0,0,0),\ (1,2,3),\ (2,1,4),\ (3,0.5,1),\ (4,4,2),\ (5,3,3),\ (6,2.5,1),\ (3.5,4). \tag{34}$$

Since $\Delta t = 1$, sample i corresponds to time i. As above, we rewrite these values:

$$\begin{aligned}(t,x) &= (0,0),\ (1,2),\ (2,1),\ (3,0.5),\ (4,4),\ (5,3),\ (6,2.5),\ (7,3.5),\\ (t,y) &= (0,0),\ (1,3),\ (2,4),\ (3,1),\ (4,2),\ (5,3),\ (6,1),\ (7,4).\end{aligned} \tag{35}$$

To define a space curve, or in other words, a true trajectory in space, we just apply our interpolation to the samples in x and y (and z) independently. That is, from the sample data of the form in Eq. 35, we define a sequence of piecewise parametric functions for x and y:

$$\begin{aligned}x(t) &= x_0(t),\ x_1(t),\ \cdots,\ x_m(t),\\ y(t) &= y_0(t),\ y_1(t),\ \cdots,\ y_m(t),\end{aligned} \tag{36}$$

where $x_i(t)$ and $y_i(t)$ are the curve segments that are defined by the basis we are using. Our space curve is

$$C(t) = C_0(t),\ C_1(t),\ \cdots,\ C_m(t), \tag{37}$$

where

$$C_i(t) = \big(x_i(t), y_i(t)\big). \tag{38}$$

All that is happening here, then, is that we are interpolating each of our spatial variables independently with respect to our parameter t. In this case, our spatial variables are x and y. We can therefore trivially extend the algorithm for Catmull-Rom cubic interpolation in Figure 2.22 to handle space curves. The refined algorithm is found in Figure 2.23. Note that to get other kinds of interpolation, we only need to insert the appropriate matrix M and change the definition of the X_i and Y_i.

A mathematical rather than algorithmic definition of each curve segment can be given as follows. Recall from Eq. 20 that for curves of degree n the result of multiplying TM gives us a set of blending functions $b_0(t), b_1(t), \cdots, b_n(t)$. A specific polynomial curve segment can thus be written as

$$p(t) = \sum_{j=0}^{n} \alpha_j\, b_j(t), \tag{39}$$

```
for i=0..n-3
    x_i(t) := T M X_i
    y_i(t) := T M Y_i
    Render((x_i(t), y_i(t)) for t ∈ [0,1])
end for
```

where

$$X_i := \begin{bmatrix} x_i \\ x_{i+1} \\ x_{i+2} \\ x_{i+3} \end{bmatrix} \text{ and } Y_i := \begin{bmatrix} y_i \\ y_{i+1} \\ y_{i+2} \\ y_{i+3} \end{bmatrix}$$

Figure 2.23. Rendering piecewise cubic Catmull-Rom space curves.

where the α_j denote the data items relevant to the curve segment. Thus any curve segment $x_i(t)$ (and analogously for $y_i(t)$) can be written as

$$x_i(t) = \sum_{j=0}^{n} \alpha_{ij} b_j(t). \tag{40}$$

The terms α_{ij}, $j = 0, \cdots, n$ denote the data items relevant to curve segment i. For example, in cubic Catmull-Rom interpolation, the data items relevant to curve segment $x_i(t)$ are

$$\alpha_{ij} = x_{i+j}. \tag{41}$$

If we plot the piecewise parametric functions $x(t)$ and $y(t)$ from the data given by Eq. 35, we get the result depicted in Figure 2.24. We can easily verify that for $x(t)$, the points $(0,0), (1,2), (2,1), \cdots$ are interpolated, and similarly for $y(t)$. Applying the above algorithm to the samples from Eq. 34 and subsequently plotting the parametric space curve $(x(t), y(t))$ gives rise to the trajectory in Figure 2.25. In this case we can verify that the points $(0,0), (2,3), (1,4), (0.5,1), \cdots$ are plotted. The time spacing between these plotted points is exactly one time unit, so it is clear that the distance that the fly travels between samples is variable.

We have finally arrived at a good solution to the modelling of trajectories. The solution can be summarised as follows. If our FlyTracker gives us a sequence of samples with respect to a time parameter t, we can dynamically create from this data a sequence of parametric polynomial curve segments $x_i(t)$, $y_i(t)$, and $z_i(t)$ using any desired piecewise interpolating polynomial basis. Each curve segment i depends only on a subset of the data sequence. Consequently, our interpolation scheme is *online* because we only need to wait for a small set of samples to arrive before the next part of the trajectory can be drawn.

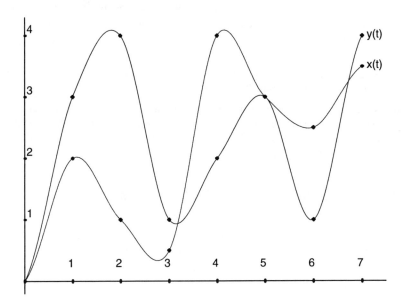

Figure 2.24. The parametric Catmull-Rom curves $x(t)$ and $y(t)$.

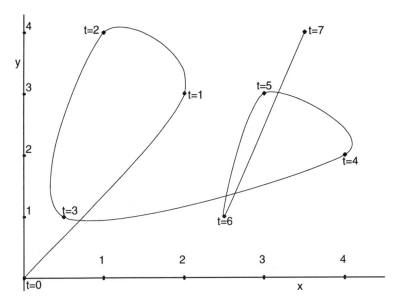

Figure 2.25. The resulting space curve $(x(t), y(t))$.

We must now address another pressing problem. Now that we have an online algorithm to define trajectories in a piecewise fashion in terms of polynomial curve segments, we must be able to evaluate polynomials very quickly to render the result in real time. We turn our attention to this problem in the next section.

Exercise 2.20. Apply four-point cubic interpolation to the above data. How does the resulting curve from the Catmull-Rom version differ?

Exercise 2.21. The interpolation bases we have considered have certain implicit assumptions regarding the number of future samples needed before the current curve segment can be drawn. Discuss these assumptions for linear, cubic Catmull-Rom, and cubic four-point interpolation.

2.5. Computational Methods for Polynomial Evaluation

At the outset of our discussion, we claimed that polynomials are easy to evaluate. It is usually the case that on a level playing field, the evaluation of a low-degree polynomial $p(t)$ is cheaper than the evaluation of a function like $\sin t$. However, the playing field is usually not level: many computers are equipped with special hardware to do some classes of arithmetic functions; often, trigonometric and exponential functions are implemented in hardware while the evaluation of polynomials is not. Recalling our discussion from the last chapter, it is therefore not necessarily true to say that the evaluation of a polynomial for a single point is faster than that for other kinds of functions. The key observation, however, is that we often need to evaluate the same polynomial at many points, and our aim will be to exploit the particular structure of polynomials to derive fast evaluation techniques that are not as applicable to other mathematical functions. There are a great many approaches to the evaluation of polynomials. We shall cover several of them in this section, including direct evaluation, Horner's rule, matrix computation, and forward differencing. In this section, we shall restrict ourselves largely to the computation of cubic polynomials, but there is little in our discussion that does not easily generalise to higher-order polynomials.

2.5.1. Matrix Computation

Recall that of the many ways of specifying a polynomial, the most convenient representation is unlikely to be the one in the power basis. A conversion from a form such as the cubic Catmull-Rom representation requires a change-of-basis matrix multiplication to convert to the power-basis representation. Thus a point on the curve for a given value of a parameter t can be computed directly from the matrix representation of a cubic parametric curve:

$$x(t) = [t^3 \ t^2 \ t^1 \ 1] \ M \begin{bmatrix} v_0 \\ v_1 \\ v_2 \\ v_3 \end{bmatrix},$$ (42)

and similarly for the y and z co-ordinates. M is a 4×4 matrix that may have zeros in some entries, and M is usually the same for each of the x, y, and z components. Recall that matrix multiplication is associative; depending on the order in which we perform multiplications, we may get different results for the number of arithmetic operations. One approach would be to take advantage of the 1 in the row vector T. First, we note that computing $[t^3 \ t^2 \ t^1 \ 1]$ itself requires 2 multiplications (abbreviated 2×).[3] Then, multiplying $[t^3 \ t^2 \ t^1 \ 1]$ and M requires 3× and 3 additions (abbreviated 3+) per column of M, giving a total of 12× and 12+. The result is a row vector multiplied by a column vector, which requires an additional 4× and 3+. The total cost of evaluating a polynomial in this manner is therefore 2× to compute the t values, and then 16× and 15+ for each of $x(t)$, $y(t)$, and $z(t)$. This is very expensive.

If we are evaluating several points on the same curve segment (as would be usual), then it is appropriate to carry out the matrix multiplication in another order to isolate the variables involving t from the constants for that curve segment. In this case, we would first premultiply M and the column vector V, giving rise to a column vector, in a preprocessing step. The cost of this step is 16× and 12+ (why?). After preprocessing, to compute the value for a particular t, we compute $[t^3 \ t^2 \ t^1 \ 1]$, requiring 2×, and we multiply it by the column vector, which requires 3× and 3+. This gives a total of 5× and 3+ for each of x, y, and z. The preprocessing cost is *amortised* over these subsequent evaluations: the initial cost can be thought of as being averaged over all subsequent evaluations of x, y, and z.

Example 2.8. Suppose we wish to evaluate eight points on one curve. Then the amortised cost of preprocessing per evaluated point is 16/8× = 2× and 12/8+ = 1.5+. The overall cost of evaluation for each point including amortised preprocessing cost is therefore 7× and 4.5+. Thus the approach based on premultiplying MV is less than half the cost of multiplying TMV for each of the eight evaluations.

As an aside, it is worthwhile to note that matrix multiplication hardware may come in handy here. Some computers have special-purpose four-element vector multipliers and 4×4 matrix multipliers available. The main reason for their existence is to perform geometric transformations of objects very quickly, but it fortuitously makes the use of cubic polynomials even more appropriate.

Exercise 2.22. Suppose we have a 4×4 matrix multiplier, as well as hardware to multiply four-element row or column vectors by a matrix. This would allow us to multiply two 4×4 matrices or multiply a 4×4 matrix and a four-element vector. Further suppose these operations work at the speed of a single scalar multiplication and that a matrix transpose operation also exists. Show how this hardware can be used to simultaneously compute four values on each of four curve segments (for a total of 16 values computed simultaneously).

---

3. These are operations on floating-point numbers as described in Chapter 5, and are fairly expensive operations on most computers. For now, it suffices to count arithmetic operations of any kind.

2.5.2. Direct Polynomial Evaluation

If we choose to perform the multiplications in Eq. 42 by first multiplying M and the column vector V corresponding to the co-ordinates of the polynomial in another basis, then the result would be

$$x(t) = [t^3 \ t^2 \ t^1 \ 1] \begin{bmatrix} a_3 \\ a_2 \\ a_1 \\ a_0 \end{bmatrix}, \tag{43}$$

which of course is nothing more than the monomial form of a cubic polynomial:

$$x(t) = a_0 + a_1 t^1 + a_2 t^2 + a_3 t^3. \tag{44}$$

Direct evaluation of $x(t)$ thus requires 3+ to sum the terms. Of course there may be fewer additions if some terms are zero. Assuming we do not have a hardware function that computes t^k in one step, each term $a_n t^n$ would require $n\times$. The total number of multiplications is $1 + 2 + 3 = 6$. The computation of $x(t)$ thus requires 3+ and 6×. So to compute a point $(x(t), y(t))$ in $\mathbf{R}^2$ requires twice this number, and three times for a point in $\mathbf{R}^3$. Once again, this is too expensive. Let us consider some simple approaches to improving this computation.

2.5.3. Horner's Rule

It is clear that a great deal of redundant evaluation is performed if we naïvely evaluate a polynomial directly from Eq. 44. It makes little sense to compute t^3 from scratch if we have already computed t^2. *Horner's rule* involves a simple rearrangement of Eq. 44 so that it is in a more computationally attractive (and therefore less visually attractive) form. We write $x(t)$ as

$$x(t) = ((a_3 t + a_2)t + a_1)t + a_0. \tag{45}$$

The cost of evaluating $x(t)$ can therefore be reduced by a factor of almost two to 3+ and 3×. However, as with direct evaluation, a preliminary matrix multiplication is required to transform the polynomial into the power-basis form (which requires 16× and 12+, as discussed earlier).

Exercise 2.23. Suppose we require m points on a curve to be computed. What is the cost per evaluation, including amortised preprocessing, of both direct evaluation and Horner's rule evaluation for cubic polynomials?

Exercise 2.24. How does the cost of evaluation grow with the degree of the polynomial using both of the above evaluation techniques?

2.5.4. Table Look-Up

As we saw in Chapter 1, the most common way to render a polynomial, or any parametric function for that matter, is to evaluate a polynomial at regularly-spaced sample points in the parameter. In this case a table look-up technique allows substantial savings in computation. For each value t_i of t needed, the value of the vector $[t_i^3 \ t_i^2 \ t_i^1 \ 1] \, M$ is computed and stored in a table. Thus we compute a table of dimensions $4 \times m$ containing the results of

$$[t_0^3 \ t_0^2 \ t_0^1 \ 1] \, M, \tag{46}$$

$$[t_1^3 \ t_1^2 \ t_1^1 \ 1] \, M,$$

$$\vdots$$

$$[t_m^3 \ t_m^2 \ t_m^1 \ 1] \, M.$$

Typically, $t_i = i \Delta t$, where Δt is a step size, though uniformly subdividing our parameter(s) is not necessary. In this case, $m = 1/\Delta t + 1$.

To compute $x(t_i)$, the vector is retrieved from the table using i as an index. Only three multiplications and three additions would then remain. Notice that this table can be used for *every* segment to be rendered using the same basis and subdivision. Therefore the cost of computing the table is amortised over all curve segments that need to be rendered.

Exercise 2.25. Suppose now that we wish evaluate s cubic curve segments and compute m points on each curve segment, all employing the same table. What is the overall amortised cost per evaluation of the table look-up technique. For concreteness, set $s = m = 8$ and compare this cost to that of the other techniques.

2.5.5. Forward Differencing Techniques

The final technique we shall discuss is much like our derivation in Chapter 1 of a line-rendering algorithm using finite differences. The approach can be extended to polynomials of arbitrary degree; this technique was well known to Charles Babbage in the 19th-Century and is called *forward differencing*. It uses the fact that the "nth difference" of a nth-degree polynomial is constant.

In Chapter 1, we wrote a differential form for a line segment in x and y as

$$\begin{aligned} x_{i+1} &= x_i + \Delta x, \\ y_{i+1} &= y_i + \Delta y. \end{aligned} \tag{47}$$

We shall only look at the x component. A more general way of writing Eq. 47 is

$$x(t + \Delta t) = x(t) + \Delta x(t), \tag{48}$$

or that

$$\Delta x(t) \;=\; x(t + \Delta t) - x(t). \tag{49}$$

This is called a the *first forward difference* of the function $x(t)$.

Forward differences can be performed iteratively, and can be used to approximate curves. For an arbitrary explicit function $f(t)$, the first forward difference is

$$\Delta f(t) \;=\; f(t + h) - f(t),$$

where h is a small positive value used in place of Δt to avoid overusing the ''Δ'' symbol. The second difference is $\Delta(\Delta f(t))$, or $\Delta^2 f(t)$:

$$\Delta^2 f(t) \;=\; \Delta f(t + h) - \Delta f(t).$$

In general, the nth forward difference is

$$\Delta^n f(t) \;=\; \Delta^{n-1} f(t + h) - \Delta^{n-1} f(t).$$

When f is a polynomial, this iteration is particularly useful. For cubics,[4]

$$
\begin{aligned}
x(t) &\;=\; at^3 + bt^2 + ct + d\,, \\
\Delta x(t) &\;=\; 3at^2 h + (3ah^2 + 2bh)t + ah^3 + bh^2 + ch\,, \\
\Delta^2 x(t) &\;=\; 6ah^2 t + 6ah^3 + 2bh^2\,, \\
\Delta^3 x(t) &\;=\; 6ah^3.
\end{aligned}
\tag{50}
$$

The last term is constant for a given h. These equations can be easily verified using Maple. Our forward differences are a function of t, and the degree decreases by one with each difference equation.

Now, suppose we wish to compute $f(t)$ at $m+1$ regularly spaced values, $t_0, t_1, \cdots, t_m$ all in $[0,1]$. Then $h = 1/m$ and $t_i = ih$, where $0 \le i \le m$. Just as we could compute a line segment iteratively, we can do the identical thing for higher-degree polynomials. In particular, if we know the value of $f(t_i)$, then we can approximate $f(t_{i+1})$ as follows:

$$f(t_{i+1}) \;=\; f(t_i) + \Delta f(t_i).$$

Unlike the rendering of line segments, the value of Δf can change. However, we can apply exactly the same argument to computing Δf, $\Delta^2 f$, etc. We can thus use the following algorithm. We initially compute

$$x(t_0), \quad \Delta x(t_0), \quad \Delta^2 x(t_0), \quad \text{and} \quad \Delta^3 x(t_0) \tag{51}$$

for a given step size h, using Eq. 50. In our case, $t_0 = 0$, and thus every term involving t in Eq. 50 disappears. Since $\Delta^3 x(t)$ is a constant independent of t, we will write it $\Delta^3 x$. Then for $i = 0$ to m we compute:

---

4. Again for notational clarity we shall write a, b, c, d for a_3, a_2, a_1, a_0, respectively.

$$x(t_{i+1}) = x(t_i) + \Delta x(t_i),$$

$$\Delta x(t_{i+1}) = \Delta x(t_i) + \Delta^2 x(t_i), \qquad (52)$$

$$\Delta^2 x(t_{i+1}) = \Delta^2 x(t_i) + \Delta^3 x.$$

For each new point only three additions are needed for each of x, y, and z, which is remarkably fast. This method has the drawback of having roundoff errors accumulating during the computations, which can make the last points computed quite inaccurate, especially if the step size h is very small.

Exercise 2.26. Suppose that in computing the initial forward differences for a cubic curve, we make an error of exactly one unit *only* in $\Delta^3 x$. Ignoring all other numerical errors that may occur, how large would the overall error be for the computation of the point $x(t_m)$? Suppose instead we make a unit error in $\Delta^2 x$ and make no errors elsewhere. What is the cumulative error in $x(t_m)$? Express your answers as a function of m. More difficult: if we render curves on a screen of integral resolution $1\,024 \times 1\,024$ with $m = 32$, how small must the units of error be to be certain that the endpoints of the curves are accurate to a pixel?

Exercise 2.27. Write a program that implements the rendering of cubic curves using direct evaluation, Horner's rule, table look-up, and forward differencing. Time their execution.

Exercise 2.28. Use Maple to compute the forward differences for fourth and fifth degree polynomials.

Exercise 2.29. What is the smallest number of arithmetic operations needed to compute the initial forward differences in Eq. 51 for $t_0 = 0$. Use this to compute the amortised cost per point on the curve, assuming m evaluations per curve.

Exercise 2.30. Use several of the previous exercises to compare the relative costs, both in preprocessing and in evaluation, of all of the polynomial-evaluation techniques presented in this section. Consider the various modalities available to you with each technique, such as step size, table size, number of curve segments to render, whether or not the change of basis varies, and so on.

2.6. Transforming Curves

2.6.1. Motivation

In many applications, it is frequently necessary to perform geometric transformations on objects. It is essential in computer graphics to be able to move an object around, to rotate it, and to change its scale. Often, the definition of a complicated object is done within a local co-ordinate system that has a conveniently-chosen origin and alignment of the co-ordinate axes. An overall scene is then composed of many objects, and the scene has its own co-ordinate system. Furthermore, in computer animation objects move in time, and their motion is often reduced to simple geometric transformations.

We saw earlier (recall Figure 2.1) that we may wish to perform interpolation with respect to one co-ordinate system, but we may wish to visualise it from another point of view. A change of co-ordinate system requires rotations, scalings, and translations. In numerical applications, the need for transformations is perhaps less obvious, but they are still often necessary. For example, our FlyTracker may report values that require rescaling, reflection or rotation after the data values have been recorded.

Given the frequent need for transformations, we must come up with ways of computing them efficiently. Explicit representations for transformations can be found in linear algebra or computer graphics textbooks and are not of concern here. It is, however, very relevant for us to consider the way in which transformations can affect a curve specification technique.

2.6.2. Formulation

Suppose we have a transformation $\mathbf{T}$ that we wish to apply to a space curve $\mathbf{C}(\mathbf{P})$ defined by a small set of points $\mathbf{P}$. We may, for example, wish first to rotate the curve $\mathbf{C}$, then to scale it in each of x, y, and z, and finally to translate it. $\mathbf{T}$ would then embody the composite transformation. It is easy to extend our discussion to the specification of curves that also work with tangents or other values, but for simplicity, we shall assume the curve-specification technique is defined only over points. The resulting curve $\mathbf{C}_1$ has a simple definition:

$$\mathbf{C}_1 = \mathbf{T}\,\mathbf{C}(\mathbf{P}). \tag{53}$$

This means that *for every* point P of interest on the curve $\mathbf{C}(\mathbf{P})$, we apply $\mathbf{T}$ to P. While this is a perfectly good mathematical definition, it can be disastrous computationally. This is because the set of points $\mathbf{P}$ defining $\mathbf{C}$ is usually much smaller than the number of points P on the curve we will need to compute. For example, for a cubic space curve, $\mathbf{P}$ consists of only four points, and it is quite likely that we will need to compute many more than four points on the resulting curve. It would be more efficient to be able to transform the data values $\mathbf{P}$ somehow rather than all the points on the curve. This intuition is captured by

$$\mathbf{C}_2 = \mathbf{C}(\mathbf{T}\,\mathbf{P}). \tag{54}$$

It would be very pleasant indeed if $\mathbf{C}_2$ were the same as $\mathbf{C}_1$. For general transformations $\mathbf{T}$, this is an exercise in wishful thinking. However, by restricting the class of transformations that we will allow and by being careful about the design of our curve-specification technique, we can actually come up with a fairly rich class of curves in which $\mathbf{C}_1 = \mathbf{C}_2$ under those transformations.

Suppose we want $\mathbf{T}$ to be an arbitrary linear transformation including an arbitrary combination of scalings and rotations but not translations. What are the constraints on our curve bases to ensure that $\mathbf{C}_1 = \mathbf{C}_2$? The answer is *none*! This will be seen in the next exercise. Suppose we wish to add translations to our list of desirable transformations. What constraints are needed now?

Let the blending functions for our n^{th}-degree curve specification technique be $b_0(t), b_1(t), \cdots, b_n(t)$, parameterised over $[0,1]$ (without loss of generality). Then the only constraint needed to ensure that $\mathbf{C}_1 = \mathbf{C}_2$ under translations is that for all $t \in [0,1]$,

$$\sum_{i=0}^{n} b_i(t) = 1. \tag{55}$$

This is called the *partition-of-unity* property. It can be visualised by drawing a vertical line denoting a value t through a graph of basis functions such as in Figure 2.15. The values of the basis functions where they meet the line must sum to one. *Every* curve-specification technique we have defined has this property, and when it does, it is said to be *co-ordinate system independent*, and *affine invariant*. This conveys the idea that (loosely speaking) the definition of a curve does not rely on a specific co-ordinate system representation. In Figure 2.26, we see an example of a single cubic space curve that has been repeatedly rotated by only rotating its control points $\mathbf{P}$. Notice that the curve retains its shape as it rotates. A curve specification that is co-ordinate system *dependent* would result in deformed curves through the rotation.

Exercise 2.31. (Easy, but requires simple linear algebra.) Let us write a curve $\mathbf{C}(t) = (x(t), y(t), z(t))$ compactly as a vector equation

$$\mathbf{C}(t) = \sum_{j=0}^{n} \mathbf{a}_j b_j(t),$$

where each observation $\mathbf{a}_j = (x_j, y_j, z_j)$ is weighted componentwise by scalar basis function $b_j(t)$. A transformation $\mathbf{T}$ is *linear* if for vectors (and points) $\mathbf{u}, \mathbf{v}$ and for scalars $a, b \in \mathbf{R}$,

$$\mathbf{T}(a\mathbf{u} + b\mathbf{v}) = a\mathbf{T}\mathbf{u} + b\mathbf{T}\mathbf{v}.$$

Rotations, scalings, and shears are linear transformations, and in fact form the class of *affine transformations*. Show that for any linear transformation $\mathbf{T}$,

$$\mathbf{T}\mathbf{C}(t) = \sum_{j=0}^{n} \mathbf{T}(\mathbf{a}_j) b_j(t).$$

While translations are not strictly linear transformations, show further that by invoking the partition-of-unity assumption for the b_j, the result holds if $\mathbf{T}$ contains a translation. Note that the proof for translations can be done independently of that for linear transformations and that the two results can then be combined.

Exercise 2.32. Prove that the Catmull-Rom basis polynomials form a partition of unity. Do the same for the four-point interpolant. Hint: analytically sum the the basis functions.

Exercise 2.33. (Difficult.) Prove that the Lagrange polynomials of degree n form a partition of unity, for each $n > 0$. See Exercise 2.10.

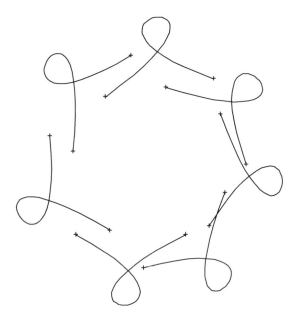

Figure 2.26. A rotated, co-ordinate system independent space curve.

There is a transformation from computer graphics that breaks the rule of co-ordinate system independence: the perspective projection. This operation involves a division by one component of a co-ordinate. With such an obvious dependence on a spatial configuration, it is not surprising that our rule is broken.

Exercise 2.34. Are there good reasons for having a curve specification that is co-ordinate system *dependent*?

Exercise 2.35. (Requires linear algebra.) Suppose we have a basis that allows the observables a_j to be either points or *tangents* to be interpolated (such as the cubic Hermite curves in the Appendix). What extra care must be used in the transformation of tangents?

2.7. An Introduction to Polynomial Surfaces

Curves are very useful for expressing trajectories, for approximating functions of a single variable, and for representing other one-dimensional objects. In computer graphics (among other areas), the representation of an object in terms of surfaces is crucial. For example, the body of a car can be modelled in part by a set of piecewise polynomial surfaces. Note that a *surface* is just that: it has no thickness, and it is not a particularly convenient representation of objects with volume. However, surface representations are often sufficient for the purposes of display. Another application is that we can use polynomial surfaces to ''zoom'' or magnify an image, a much different application from the design of a car.

We have seen examples of surfaces in Chapter 1 such as the sphere. The parametric definition of the sphere is actually a simple extension of the definition of a circle. Analogously, a large class of polynomial surfaces can be built directly from polynomial curves. We cannot express all polynomial surfaces using the form to be described, but the class is sufficiently rich for many applications. In particular, this class is the *tensor-product* form for polynomial surfaces. Suppose we wish to define a degree-n polynomial surface $p(s,t)$ defined now over two parameters s and t both varying on $[0,1]$. The idea of this form is that if we vary one parameter but leave the other constant, the result is an nth-degree isoparametric curve. If both s and t are allowed to vary, then the result can in fact be a very complicated surface. The form for P is

$$p(s,t) \;=\; T M V M^+ S^+, \tag{56}$$

where M is our familiar $(n+1) \times (n+1)$ change-of-basis matrix with transpose M^+, S is a row vector $[s^n \; s^{n-1} \; \cdots \; s^1 \; 1]$ of the power-basis in parameter s, and similarly $T = [t^n \; t^{n-1} \; \cdots \; t^1 \; 1]$ is a row vector of the power-basis in parameter t. The $(n+1) \times (n+1)$ matrix V is a set of data values to be interpolated. We need such a polynomial surface for each of x, y, and z.

In basis function form, this representation shows the separation of a bivariate surface $p(s,t)$ into the product of two univariate functions. For a degree n bivariate surface, we define n^2 data values α_{ij} (as in V above) and univariate basis functions b_k, $k = 0, 1, \cdots, n$, and put this all together as follows:

$$p(s,t) \;=\; \sum_{i=0}^{n} \sum_{j=0}^{n} \alpha_{ij}\, b_i(s)\, b_j(t). \tag{57}$$

While the notation looks rather intimidating, the idea is actually very simple. Let us consider the familiar cubic case, in which $n = 3$. For each component x, y and z think of V (or the α_{ij}) as being a uniformly-spaced 4×4 grid of *height* values, where the parameters s and t define a horizontal floor plane. Then in this case $p(s,t)$ defines a continuous surface hovering over the floor that smoothly blends all 16 points. A single bicubic "bump" over 16 points is illustrated in Figure 2.27. The resulting surface is called a *bicubic surface*; the piece of the surface corresponding to a fixed parametric domain of, say, $[0,1] \times [0,1]$ is called a *bicubic patch*. If $n = 1$, the result is called a *bilinear patch*, or *bilinear interpolation*. Linear interpolation in each dimension is widely used in many applications.

Just as we were able to create complex curves by stitching together pieces of polynomial curves, we can do the same thing with surfaces. Figure 2.28 illustrates the effect of putting together 32 bicubic patches in a not-very-random fashion. The teapot data was originally input using a light pen and a real teapot. The use of bicubic patches derived from a tensor-product form is an essential component in computer graphics modelling systems. All of the techniques we discussed above for the computation of curves have a two-dimensional analogue.

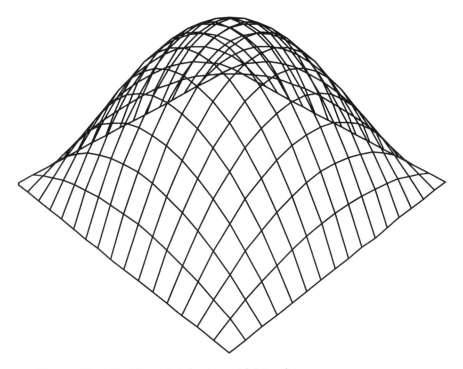

Figure 2.27. A bicubic patch defined over 16 data points.

Appendix A: Computing the Change-of-Basis Matrix

It is easy to design our own curve-specification techniques. In this appendix, we consider how we can come up with ways of specifying curves that are more suitable for some applications.

We have been writing a general polynomial of degree n as

$$p(t) = TA, \qquad (58)$$

where $T = [t^n \ t^{n-1} \ \cdots \ t \ 1]$ is a row vector containing the $n+1$ polynomials in t in monomial form, and $A = [a_n \ a_{n-1} \ \cdots \ a_0]^+$ is a column vector of coefficients. The corresponding space curve $\mathbf{p}(t) = (x(t), y(t), z(t))$ would have an identical definition in each component. For short, we could write

$$\mathbf{p}(t) = T\mathbf{A}.$$

Since the parametric formulation operates independently over each component, we can derive the change of basis with respect to a single component. It thus suffices to talk about a curve $p(t)$ rather than a space curve $\mathbf{p}(t)$.

We saw earlier that modifying the behaviour of a curve p is difficult, if all we are allowed to do is to change the values of the coefficients. Thus we worked

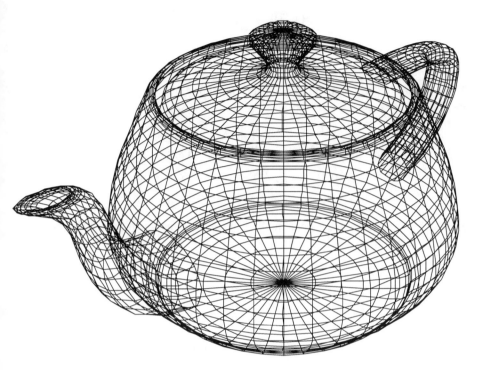

Figure 2.28. A teapot composed of 32 bicubic patches and 306 data points.

with other curve-specification techniques such as the Catmull-Rom and four-point interpolant, which allowed us to rewrite $p(t)$ as

$$p(t) \;=\; T M V,$$

such that

$$A \;=\; M V.$$

In this case, V represents a column vector that contains $n+1$ "observations" regarding the desired behaviour of the curve, and M is a change-of-basis matrix. Finally, we also saw that in effect, this representation amounts to defining the polynomial $p(t)$ as a point V in the function space with polynomial basis functions (or blending functions) given by multiplying $T M$. In this appendix, we consider how to derive M for various curve representations.

Suppose we wish to derive the change-of-basis matrix, M_I for four-point interpolation, as was given in Eq. 24. We needed to define four constraints, and they were that the polynomial $p(t)$ must interpolate four values v_0, v_1, v_2, v_3 on the interval $[0,1]$ as follows:

$$p(0) \;=\; v_0,$$

$$p(1/3) = v_1,$$
$$p(2/3) = v_2,$$
$$p(1) = v_3.$$

These constraints say something quite specific about the behaviour of $p(t)$; from Eq. 58 they say that:

$$p(0) = [0\ 0\ 0\ 1]\,A = v_0,$$
$$p(1/3) = [(1/3)^3\ (1/3)^2\ (1/3)\ 1]\,A = v_1,$$
$$p(2/3) = [(2/3)^3\ (2/3)^2\ (2/3)\ 1]\,A = v_2,$$
$$p(1) = [(1)^3\ (1)^2\ (1)\ 1]\,A = v_3.$$

Since these conditions must hold simultaneously, this mess can be written more neatly in matrix form:

$$B\,A = V.$$

Here, $V = [v_0\ v_1\ v_2\ v_3]^+$ is a column vector corresponding to the values to be interpolated, and B is the 4×4 matrix corresponding to the above row vectors:

$$B = \begin{bmatrix} 0 & 0 & 0 & 1 \\ 1/27 & 1/9 & 1/3 & 1 \\ 8/27 & 4/9 & 2/3 & 1 \\ 1 & 1 & 1 & 1 \end{bmatrix}.$$

We therefore have four equations in four unknowns, namely the coefficients A. To solve for A, we note that

$$A = B^{-1}\,V.$$

Hence we have an immediate representation for our change-of-basis matrix M:

$$M_I = B^{-1},$$

if the inverse exists. In this specific case, the inverse does indeed exist, and we have the matrix used in Eq. 24 above:

$$M_I = \begin{bmatrix} -9/2 & 27/2 & -27/2 & 9/2 \\ 9 & -45/2 & 18 & -9/2 \\ -11/2 & 9 & -9/2 & 1 \\ 1 & 0 & 0 & 0 \end{bmatrix}.$$

As we said above, this coincides with the cubic Lagrange interpolating polynomial on four equally spaced points over [0,1]. In summary, then, the goal is to collect a series of independent constraints on the behaviour of the curve in matrix form, and then to invert that matrix to get our change of basis.

Another example would be instructive. The set of constraints above were all defined with respect to interpolating points in space. Sometimes it is useful to have *mixed* constraints in which we write down equations that specify interpolation and other properties. The following curve constraints induce one of the best known cubic curve formulations.

Suppose we wish to define a cubic curve by specifying that the endpoints of the curve over [0,1] must be interpolated, and that the *tangent* to the curve at 0 and 1 must also be interpolated. To put things into context, suppose our tracker is able to report not just the position of the fly but its *velocity* as well. In this case it would be useful to be able to interpolate position and velocity simultaneously. This is easy to do. Mathematically, the constraints on the corresponding space curve $\mathbf{p}(t)$ are:

$$\mathbf{p}(0) = V_0,$$
$$\mathbf{p}'(0) = \mathbf{T}_0,$$
$$\mathbf{p}(1) = V_1,$$
$$\mathbf{p}'(1) = \mathbf{T}_1,$$

for points V_0, V_1 and tangents $\mathbf{T}_0, \mathbf{T}_1$. As usual, it suffices to look at $\mathbf{p}(t)$ componentwise. Let the polynomial curve in one component be $p(t)$, and let p' denote the derivative of p with respect to t. Furthermore, we denote the relevant values of V_0, etc., by the corresponding lower case letter such as v_0. If

$$p(t) = [t^3 \ t^2 \ t \ 1]A,$$

then basic calculus says that

$$p'(t) = [3t^2 \ 2t \ 1 \ 0]A.$$

We can therefore write down a set of mixed constraints as follows.

$$p(0) = [0 \ 0 \ 0 \ 1]A = v_0.$$
$$p'(0) = [0 \ 0 \ 1 \ 0]A = t_0.$$
$$p(1) = [1 \ 1 \ 1 \ 1]A = v_1.$$
$$p'(1) = [3 \ 2 \ 1 \ 0]A = t_1.$$

Once again, we can write this down in matrix form:

$$\begin{bmatrix} 0 & 0 & 0 & 1 \\ 0 & 0 & 1 & 0 \\ 1 & 1 & 1 & 1 \\ 3 & 2 & 1 & 0 \end{bmatrix} A = \begin{bmatrix} v_0 \\ t_0 \\ v_1 \\ t_1 \end{bmatrix}.$$

If we invert the left-hand matrix, we get the following matrix:

$$M_H = \begin{bmatrix} 2 & 1 & -2 & 1 \\ -3 & -2 & 3 & -1 \\ 0 & 1 & 0 & 0 \\ 1 & 0 & 0 & 0 \end{bmatrix}.$$

This form of interpolation is called *cubic Hermite interpolation*. Multiplying $[t^3 \ t^2 \ t \ 1] M_H$ gives us the *cubic Hermite blending functions*:

$$h_0(t) = 2t^3 - 3t^2 + 1,$$

$$h_1(t) = t^3 - 2t^2 + t,$$

$$h_2(t) = -2t^3 + 3t^2,$$

$$h_3(t) = t^3 - t^2.$$

These functions are plotted in Figure 2.29. As with other blending functions we have seen, notice that at $t = 0,1$, the value of three of the blending functions is zero and the other has a value of one. Once again, therefore, this basis is in cardinal form. Also notice that the blending function h_1 for T_0 is always positive, whereas the blending function h_3 for T_1 is always negative.

There is a direct relationship between cubic Catmull-Rom and cubic Hermite interpolation. Recall that for a sequence A, B, C, D of points, the resulting cubic Catmull-Rom polynomial interpolates B and C. In fact, the *tangents* to the curve at B and C are related to A,C and B,D, respectively. Specifically, the equivalent cubic Hermite curve is got by:

$$\begin{bmatrix} V_0 \\ T_0 \\ V_1 \\ T_1 \end{bmatrix} = \begin{bmatrix} B \\ \dfrac{C-A}{2} \\ C \\ \dfrac{D-B}{2} \end{bmatrix}. \tag{59}$$

This means that the derivative of a Catmull-Rom curve at $t = 0$ is directly related to the slope of the line segment AC, and analogously at $t = 1$. Whether or not this is acceptable depends on the application, but if the input data is nothing but points to be interpolated (rather than the mixture of tangents and points as in the Hermite case), then *some* constraints on tangency at the joints are necessary in order to get smoothness between curve segments.

Exercise 2.36. Easy: write down the change-of-basis matrix $M_{CR \to H}$ that takes us from the cubic Catmull-Rom form to the Hermite form. That is, derive the matrix $M_{CR \to H}$ such that

$$M_{CR} = M_H M_{CR \to H},$$

where M_{CR} is the Catmull-Rom basis defined in Eq. 26.

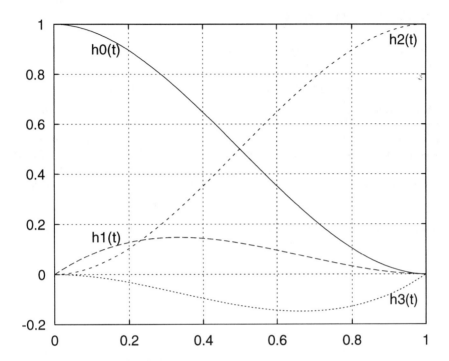

Figure 2.29. The cubic Hermite blending functions.

Exercise 2.37. Harder: write down the change-of-basis matrix $M_{H \to CR}$ that takes us from the cubic Hermite form to the Catmull-Rom form.

Exercise 2.38. How should cubic Hermite polynomial curve segments be stitched together?

Exercise 2.39. Use Maple to derive a curve-specification technique that, in addition to the interpolating both the position and velocity at $t = 0,1$, also interpolates the acceleration at $t = 0,1$. Of what degree must the resulting curves be? What are the dimensions of the change-of-basis matrix? Write down and plot the blending functions for this class of curves. What is a suitable name for this class?

Exercise 2.40. Consider a cubic basis that satisfies the following constraints: given points A, B, C, D, produce a polynomial $p(t)$ with $t \in [0,1]$ such that

$$p(0) = A, \quad p(1) = D, \quad p'(0) = 3(B - A), \quad p'(1) = 3(D - C).$$

Derive the change-of-basis matrix and blending functions for this form. The resulting polynomial p is called a cubic *Bezier* curve.

Supplementary Exercises

Exercise 2.41. Write a Maple routine called **PiecewiseLagrange** that has the following invocation sequence:

PiecewiseLagrange(f, d, nseg, nsamp)

where **f** is a function of one variable, **d** is a domain of the form **[a,b]**, **nseg** is the number of piecewise continuous Lagrange polynomials to be stitched together, and **nsamp** is the number of samples of **f** to be taken within each segment. The **nsamp** samples in each segment are to be equally spaced over the subinterval of **[a,b]** corresponding to that curve segment. The goal of this routine is to plot both the function **f(x)** and the approximating piecewise Lagrange polynomial over **[a,b]**. For example,

PiecewiseLagrange(sin(x), [0,2*Pi], 8,2);

would produce a plot of $\sin x$ over $[0,2\pi]$ as well as a piecewise linear interpolation consisting of eight segments. On the other hand,

PiecewiseLagrange(sin(x), [0,2*Pi], 1,8);

would plot a single polynomial segment of degree seven (since there are eight points to be interpolated), together with $\sin x$. There is a function called **ApproxLagrange** in Chapter 5 that is somewhat helpful, but it is not quite in the right form. Once you modify it, you can use it as a subroutine. Note: coming up with an interpolating polynomial for each segment is easy; putting an overall curve together from these segments so that Maple can plot it is quite difficult. You should only invoke a single **plot** instruction to view the composite of the original function together with the approximation. Note that **nsamp** is an integer and cannot be smaller than 2, while **nseg** cannot be smaller than 1.

Once you have tested out your routines, examine the performance of your interpolation package on the following functions:

```
PiecewiseLagrange(cos(x)*sin(x),[-3,1], 4,d);      (d=2,3,4)
PiecewiseLagrange(sin(x)/x,[0.1,10], 9,d);         (d=2,3,4)
PiecewiseLagrange(cos(x)^k,[-2,2], n,d);           (vary n,d,k)
```

Some really tough tests are:

- $\sin(1/x)$ over a interval $[\varepsilon,0.3]$, where ε is close to zero (e.g., 0.05). This is an excellent example of the effects of undersampling. You'll need quite a few segments to handle this (Maple has trouble plotting it as well).

- $x^{1/x}$ over a variety of intervals, segments and degrees.

- To see ringing (see next chapter), try x^x on an interval like [8,20]. Since this function increases so quickly, you need higher degree segments, but this causes wiggling. Try it!

Chapter 3
Approximation and Sampling

As we saw in the last chapter, the mere fact that we can interpolate some samples taken from a function does not necessarily mean that we have approximated the function well. In fact, the constraint of interpolation can make our "approximation" arbitrarily bad almost everywhere except at the points we are interpolating. Approximating a function—perhaps even interpolating no points on the curve—is sometimes preferable to interpolation. There is a vast array of approximation techniques, and we cannot hope to cover many of them in depth. In this chapter we shall introduce approximation as motivated by the discrete sampling and reconstruction of functions. The topics to be discussed include:

- how many samples are needed to approximate an unknown function?
- uniform vs. nonuniform sampling.
- coping with noisy input data.
- convolution and filtering.
- applications.
- computational issues.

This chapter, then, introduces some of the main concepts of *signal processing* by way of considering the problems introduced when interpolating unreliable data. Virtually all experimental data that is converted to digital form has error in it. We shall continue with our motivating problem of tracking a fly's trajectory from measured data, but we shall consider other applications as well. Our discussion in Chapter 5 will be devoted to approximation strategies involving various classes of series.

3.1. Problems with Interpolation

When we considered the use of Lagrange interpolation to compute a single global polynomial to model a fly's trajectory, we discovered that our approach was flawed: the resulting polynomial was of high-degree and unwieldy, it was not an online or dynamic approach, and the arc length of successive computations differed wildly. For this reason, we suggested the use of piecewise, low-degree polynomial approaches over a variety of bases, including Lagrange, Catmull-Rom, and Hermite. There are many times, however, when the constraint of interpolation leads us into difficulty. We introduced some of these difficulties as exercises in the last chapter. We briefly consider some graphical examples of the problems inherent to interpolation and sampling. Most of these problems will be considered in greater detail later in the chapter.

3.1.1. Ringing

A large difference in value between two samples can cause the interpolated trajectory (or approximated function) to overshoot the jump and then subsequently stabilise. In the signal processing literature, this phenomenon is called *ringing*. A good example arises when trying to interpolate samples derived from a discontinuous *step* function. A sample step function would be:

$$f(x) \;=\; \begin{cases} 0 & \text{if } x < 4. \\ 1 & \text{if } x \geq 4. \end{cases} \tag{1}$$

It is plotted together with interpolating Lagrange and piecewise cubic Catmull-Rom polynomials in Figure 3.1. Remember that our interpolation operates from knowledge about a set of samples: the actual function itself is not available, so the location of the discontinuity is unknown. Observe that the Lagrange polynomial has worse ringing behaviour. In fact, this behaviour worsens with increasing degree. In general, however, all interpolating formulations having degree higher than one will ring in the vicinity of the abrupt change. It is also easy to cause ringing with a continuous function: see Figure 2.4 in the last chapter, for example.

Exercise 3.1. How could you cause piecewise linear interpolation to exhibit ringing?

Exercise 3.2. Plot the Lagrange basis functions of degree nine (or just look at the basis functions for the four-point interpolant from the last chapter). Compare them to the basis functions for cubic Catmull-Rom interpolation. Look for clues in the basis functions that would suggest different ringing behaviour between these formulations.

Of course, if we knew beforehand that we would be modelling a step function, we could isolate the discontinuity, and then we could employ a piecewise linear interpolant, but such foresight is rarely available to us in real applications. In some cases, however, a *discontinuity* or *variational analysis* is possible. In our case, the function itself is unavailable, since the flight of our fly is presumably not

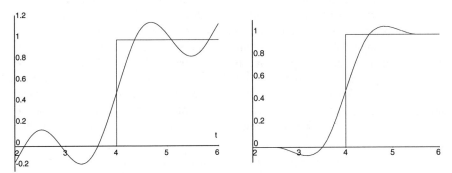

Figure 3.1.   Approximation to a step function plotted in the interval [2,6]. Left: a portion of a 9th-degree Lagrange polynomial. Right: two piecewise cubic Catmull-Rom segments.

governed by some algebraic function. If we model a fly's trajectory using an interpolating formulation, unless we politely ask the fly not to change its course abruptly, and the fly generously consents to doing so, we are certain to overshoot the actual trajectory occasionally.

It is useful to consider a natural analogue for ringing. A real *spline* is a strip of wood or metal that is warped by applying pressure at various points. We can think of our basis functions as applying the forces, and the resulting curve as being a wooden strip that bends due to these forces. In an interpolating formulation such as Catmull-Rom or Lagrange, the strip comes into contact with, and is clamped at, certain points, specifically the knots at which points are to be interpolated. To prevent fracture or buckling of the strip, its curvature must be bounded, depending on its material properties. To conform to a set of samples from a step function without fracturing, a real strip would have to dip below zero, and then rise quickly to the value one; the strip then goes beyond one and subsequently recovers. A physical spline is probably closer to the cubic Catmull-Rom segments than to the Lagrange interpolation. In fact, the piecewise cubic *natural splines* are an extremely good model for physical splines.[1]

While ringing behaviour is mildly unacceptable for describing the flight of a fly, it can be disastrous in other applications. Suppose the samples instead represent the dosage of some drug that must be applied intravenously to a patient at specific time instants, and that our job is to create a continuous regulation of the drug over a time interval. Then our interpolation above would suggest that at certain times we would have to dispense negative quantities of the drug and that at others, we could give the patient an overdose!

---

1. See J. Hoschek and D. Lasser (translated by L.L. Schumaker), *Fundamentals of Computer-Aided Geometric Design*, A.K. Peters, Wellesley, MA, 1993.

Another application in which disastrous (but not tragic) results can occur arises in the reconstruction of a continuous musical signal from a set of discrete samples. A compact disk, for example, contains discrete samples defining the frequency and amplitude (or magnitude) of the musical signal at time t. A musical signal that closely approximates a step function is the striking of a tightly tuned drum (or the pizzicato snap of a stringed instrument). If we were to use an interpolating formulation as above, we could produce a distorted output voltage that could greatly exceed the capacity of the amplification system. The result, if the amplifier does not have a circuit breaker, could be the destruction of the speaker system.

The most graphic example of ringing can be seen in Figure 3.2. Here, the "step" functions are given by the image on the left, which contains sharp transitions from white to black. The image on the right is an artificially low-quality encoding of the original using the JPEG standard. This encoding is based on a piecewise, discrete cosine transform (see Chapter 5); reconstruction (i.e., interpolation) is based on an inverse cosine transform, which just means that it uses sinusoids as basis functions. Observe the ringing artifacts, which are visible as grey blotches on the white regions. Notice as well that the artifacts are *anisotropic*, meaning that the effect varies with orientation of the line segments. The extra jagginess in the oblique line also causes the encoding to behave worse. In fact, echoes of parts of the original line segment should be visible. To be fair to JPEG, the image on the right was created using an extremely low-quality setting. Using normal quality settings, a JPEG encoding is usually excellent.

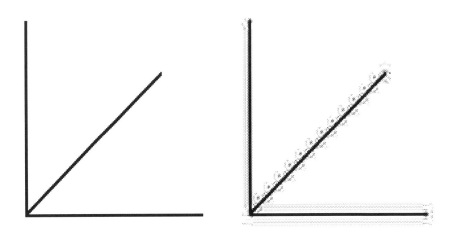

Figure 3.2. Visual evidence of ringing. Left: the original image. Right: a JPEG encoding using a very low-quality setting.

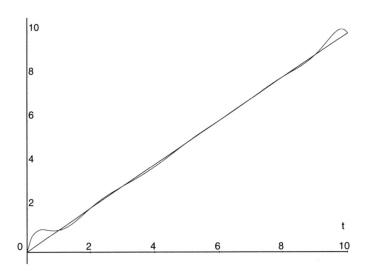

Figure 3.3.   Two sample trajectories: the straight line corresponds to perfect data,
while the curve corresponds to displacing a single point by 2.5%.

3.1.2. Noise

Just as discontinuities or abrupt changes in sample values can cause large-scale
ringing artifacts, small-scale errors in the samples can have an amplified effect on
the interpolated trajectory. For example, Figure 3.3 gives the results of plotting
two very similar sets of sample data. The straight line is the result of interpolating

$$(0,0), (1,1), (2,2), \cdots, (10,10)$$

using the Lagrange formulation, while the wavy trajectory is the result of
interpolating a set of samples that is identical to the above, except that $(4,4)$ is
replaced by $(4,3.9)$. Figure 3.4 depicts a similar result when each point except
$(0,0)$ and $(10,10)$ is displaced by a small positive or negative random value.

3.1.3. Undersampling

We have generally assumed that our samples (x,y,z,t) are equally spaced in the
parameter t. A great many applications make this assumption, but the following
two problems immediately confront us:

- How many samples do we need to take before we can be certain that we
  can construct an accurate representation for a function or trajectory?
- What spacing should we choose for t?

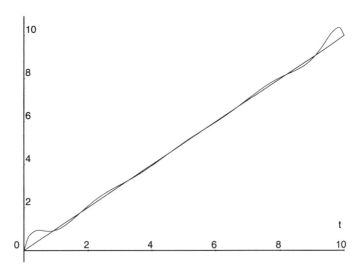

Figure 3.4.   Two sample trajectories: the straight line again corresponds to perfect
data, while the curve corresponds to randomly displacing the x values
of the nine interior points.

Unfortunately, in many cases, we have control over neither the number of samples
nor their spacing, but the wealth of theory that addresses these two related
problems spells out the set of functions that can be approximated when given a
certain sampling rate. Suppose, for example, our fly actually travels in an
oscillating trajectory given by $y = \cos t$ for $t \in [-4\pi, 4\pi]$, but that our bargain-
basement tracker can only emit samples every 2π seconds. In this case, our
tracker would give the values

$$(-4\pi, 1), (-2\pi, 1), (0, 1), (2\pi, 1), (4\pi, 1). \tag{2}$$

By any reckoning, this is a straight line, and it satisfies our constraint that it must
interpolate the sample points on the curve. It is hardly a good approximation to
the original curve. The two trajectories are modelled in Figure 3.5. Notice as
well that if we were to change the time at which we begin to sample, while
keeping the same sample spacing, we would still get a straight line, but for a
different value of y. Any way we take four uniformly spaced samples, we are
clearly *undersampling* the actual trajectory of the fly. Later, we shall consider the
problem of how many samples are sufficient for us to be confident that we can
reproduce accurately the fly's actual trajectory. We shall see that in fact we often
cannot be confident and must adopt schemes to improve our chances.

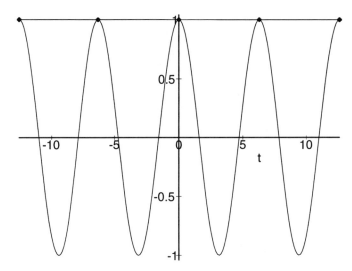

Figure 3.5.   The cosine trajectory and the linear trajectory due to undersampling.

3.1.4. Divergence

It seems obvious that if we do not take enough samples, we will not be able to construct a reasonable approximation to a function. It is also at least understandable that noisy or abruptly changing values will cause more unstable interpolations. What is more surprising is that even if we take many equally spaced samples of a continuous function, some approaches to constructing an interpolating curve can actually *diverge* from the result as the number of samples increases. Note that this is not inconsistent with our claim that an arbitrary continuous function can be approximated by a polynomial of sufficiently high degree. It's simply that our choice of polynomial has to be a particularly careful one in some cases. We shall return to this discussion below, but it is useful to look at an example.

Consider the function first analysed by Runge in the early 1900s:

$$R(x) \;=\; \frac{1}{1\,+\,25x^2}. \tag{3}$$

Most of us have seen far more intimidating functions that this. A graph of this function on the interval $[-1,1]$ does little to frighten us either, as we can see in Figure 3.6. However, if we consider successive approximations using Lagrange interpolation, we get very poor results. In particular, as the number of samples increases, our approximation of the function for $|x| > 0.7$ gets increasingly bad. Figure 3.7 illustrates the behaviour of Lagrange interpolation on 6, 11, and 16 uniformly spaced samples of R.

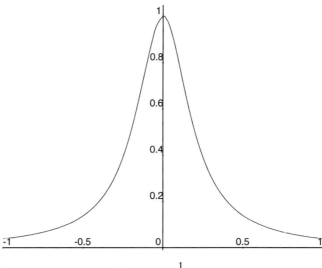

Figure 3.6. Runge's function: $R(x) = \dfrac{1}{1 + 25x^2}$.

Perhaps this poor behaviour can again be blamed on the unwise use of high-degree polynomial interpolations. It is true that piecewise cubic Catmull-Rom interpolation yields good results on seven uniformly spaced samples (see Figure 3.8). The main point here, however, is that we must be extremely careful about the manner in which we *reconstruct* a function from a set of samples that has been handed to us. We shall revisit this problem below.

3.1.5. Summary

This section has illustrated some of the more severe problems that can result when we ask too much of interpolation. We have seen that highly variable or noisy data, undersampling, and poor reconstruction strategies can cause approaches based on interpolation to behave very poorly. For the remainder of this chapter, we shall use these problems as motivation for other classes of approximation techniques.

3.2. Types of Approximation

Let us suppose our primary goal is to compute a smooth trajectory that is close to a set of samples. Since we have seen that noisy or variable data can cause problems if we use interpolation, then perhaps we should forego interpolation altogether and instead settle for a smooth, non-ringing curve that is acceptably

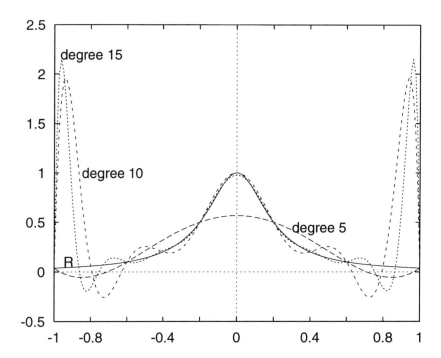

Figure 3.7.   Approximation of Runge's function using Lagrange interpolation of various degrees.

close to the data. There are many ways of defining "close." One way is to compute the distance between a distinguished point and the closest approach of the curve to that point. We could then sum such distances for a set of points to get a very coarse measure of closeness of *fit* of the curve to the data. This measure should also take into account the degree of "wobble" in the curve so as to penalise artifacts such as ringing.

An approach to this problem is to look at it as a problem in optimisation. Assuming we have an objective way of measuring the quality of fit of a curve to a set of samples, we can formulate an expression whose minimisation gives us the best possible curve from a possibly very large set of curves. If we define our quality metric appropriately, then the method of *least-squares approximation* can be employed. In general a deep theory called the *calculus of variations* is required to analyse the underlying structure of approaches to optimisation. For our purposes, these approaches are not particularly appealing for several reasons. First, they are computationally subtle. Second, the behaviour of the curve actually chosen would not be available until after a potentially long computation, making it inapplicable to our real-time trajectory model. Third, a strategy based on

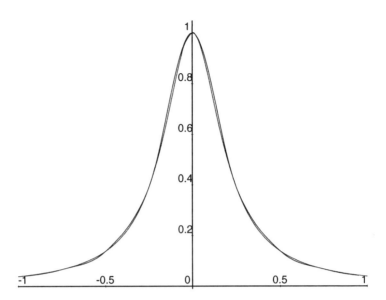

Figure 3.8. Cubic Catmull-Rom approximation to Runge's function using seven samples.

computing an overall or *global* optimum with respect to some objective function is not necessarily the most desirable solution. Global optimisation approaches are, however, of great importance to many aspects of science and are widely used.

Another class of approaches is one in which we have more direct *local control* over the result. The size of this class is as vast as those based on optimisation. In fact relationships exist between various classes of solutions; some solutions that have desirable local control can also be optimal in the sense of minimising the variation of *curvature*, or the sharpness with which a curve bends. While we cannot prove this result here, the piecewise cubic B-splines that we shall soon discuss have this property.

If we know that our sample data will have problems such as noise in it, then yet another (vast) set of approaches favours processing the data beforehand to remove or at least reduce the effect of these problems. For example, scanning devices for medical imaging rarely produce "clean" data, but the distortion introduced by the equipment has a characteristic behaviour that can be largely removed from the input, leaving more meaningful data behind. Such approaches are typically based on *filtering*. We shall discuss some efficient, commonly employed discrete filtering techniques and compare them to other approximation techniques.

3.3. Approximation Using Uniform Cubic B-Splines

An approach to dealing with problem of messy data is to use piecewise polynomial *approximation* rather than interpolation. These approaches are important in many other areas as well. There is a huge literature on this topic, but by way of example, we consider perhaps the most popular piecewise cubic curve formulation: the *uniform cubic B-spline*. Just as for the other cubic formulations we discussed in the previous chapter, this formulation operates piecewise using four observations to define a cubic polynomial for each curve segment $\mathbf{p}_i$ (recall that the bold notation means this is a parametric space curve). These four observations are positions

$$P_i, P_{i+1}, P_{i+2}, P_{i+3} \in \mathbf{R}^n \tag{4}$$

in space, just as for the cubic Catmull-Rom and Lagrange bases. Our B-spline basis is *uniform* in the sense that the knots are equally spaced. Remember that a *knot* is a value of the parameter (i.e., in the domain) corresponding to a position at which we wish to affect the behaviour of the curve. It often refers to a point in parametric space at which curve segments join. Thus four uniformly-spaced points in a parameter over [0,1] would have knot values 0, 1/3, 2/3, and 1. The *B* in *B-spline* originally stood for *basic*, but it could just as easily stand for *basis* or *blending*. All of the cubic forms we have considered can be expressed as higher-degree B-splines. We shall focus on the class of cubic B-splines, which are distinguished by having second-derivative parametric continuity at the joints between curve segments. None of the other cubic formulations we have discussed have this property. The constraints satisfied by a cubic B-spline are that:

- The endpoints of the curve segments must agree. This does not mean that the curve must interpolate the points, only that the relevant points must be shared between curve segments. So if $\mathbf{p}_i(t)$ and $\mathbf{p}_{i+1}(t)$ are two adjacent curve segments with parameter $t \in [0,1]$, then $\mathbf{p}_i(1) = \mathbf{p}_{i+1}(0)$.

- The tangent to the curve at the joints between curve segments must agree. More formally, if $\dot{\mathbf{p}}_i(t)$ and $\dot{\mathbf{p}}_{i+1}(t)$ are the componentwise first derivatives with respect to t of $\mathbf{p}_i(t)$ and $\mathbf{p}_{i+1}(t)$, then $\dot{\mathbf{p}}_i(1) = \dot{\mathbf{p}}_{i+1}(0)$.

- The second derivative to the curve at the joints between curve segments must similarly agree.

We saw that when we performed piecewise Lagrange interpolation, we could easily ensure that endpoints of each curve segment matched. In this case, the resulting space curve $\mathbf{p}$ would be *continuous*. However, there was no mechanism for constraining the derivative at the end of each segment. Thus it is very likely that $\dot{\mathbf{p}}_i$ would be different from $\dot{\mathbf{p}}_{i+1}$ at the point they share. The derivative $\dot{\mathbf{p}}$ of the space curve would be *discontinuous*, resulting in cusps at shared points. Both the cubic Catmull-Rom and the cubic Hermite forms ensure that the first derivative is always continuous, but there is no such constraint on the second derivative. The added *degree* of continuity inherent in the cubic B-spline has a significant effect on the resulting curves, as we shall see.

The formulation of a change-of-basis matrix for cubic B-splines can be arrived at from the above constraints, but more commonly B-splines of arbitrary degree n are defined recursively from those of degree $n-1$. This construction is somewhat intricate for our purposes, so we shall just state the following basis matrix for cubic B-splines:

$$M_b = \frac{1}{6} \begin{bmatrix} -1 & 3 & -3 & 1 \\ 3 & -6 & 3 & 0 \\ -3 & 0 & 3 & 0 \\ 1 & 4 & 1 & 0 \end{bmatrix} \tag{5}$$

Exercise 3.3. Solving for this matrix will require more ingenuity than the techniques described in Appendix C of the last chapter. Formulate a list of equations corresponding to the continuity constraints for the uniform cubic B-spline. Do you have enough equations to solve for M_b? If not, what extra constraints might you want? Hint: partition of unity.

Putting this matrix into our general form for polynomial space curves gives

$$\mathbf{p}_i(t) = [t^3 \quad t^2 \quad t \quad 1] M_b \begin{bmatrix} P_i \\ P_{i+1} \\ P_{i+2} \\ P_{i+3} \end{bmatrix} \tag{6}$$

$$= BP.$$

In this case, recall that the space curve $\mathbf{p}_i$ actually corresponds to n polynomial curves assuming the $P_j \in \mathbf{R}^n$: one for each component x, y, z, $\cdots$. The symbol B denotes the row vector $[b_0(t) \quad b_1(t) \quad b_2(t) \quad b_3(t)]$ defining the uniform cubic B-spline polynomial blending functions:

$$b_0(t) = -\frac{1}{6}t^3 + \frac{1}{2}t^2 - \frac{1}{2}t + \frac{1}{6},$$

$$b_1(t) = \frac{1}{2}t^3 - t^2 + \frac{2}{3},$$

$$b_2(t) = -\frac{1}{2}t^3 + \frac{1}{2}t^2 - \frac{1}{2}t + \frac{1}{6}, \tag{7}$$

$$b_3(t) = \frac{1}{6}t^3.$$

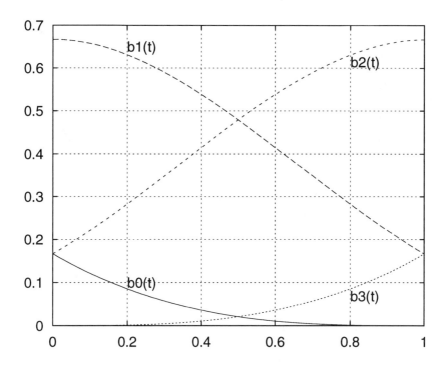

Figure 3.9.   Cubic B-spline blending functions.

These functions are plotted in Figure 3.9, and have a few notable features. First, observe that unlike any of the other cubic blending functions we have discussed, *none* of the curves is one when the others are zero. Indeed, the value along any of the curves never exceeds 2/3 (why?). Thus we cannot expect this curve to interpolate any specified points (except accidentally). Second, the curves are always positive in value. Third, although this is not entirely easy to see, if we sum the values of the blending functions for any value of $t \in [0,1]$, we get the value 1. Hence these blending functions have the *partition-of-unity* property.

Exercise 3.4. Verify that the cubic B-spline blending functions are a partition of unity.

Our presentation of B-splines is very limited. B-splines have a rich theory that admits multiple knots and nonuniform knot spacing. Our representation of cubic B-splines in matrix form depends on uniform knot spacing and a uniform parameterisation of all polynomials over a specific interval, in our case [0,1].

The use of cubic B-splines in a piecewise setting is exactly the same as that for the cubic Catmull-Rom form, and it is given by Eq. 6 above. Note that curve

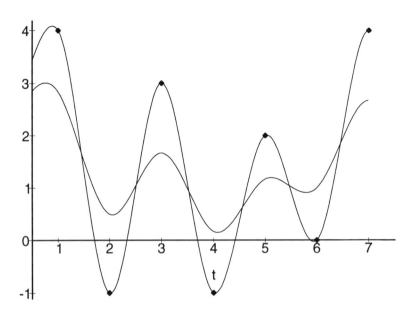

Figure 3.10. Cubic Catmull-Rom interpolation vs. cubic B-spline approximation of
the same data. Which is which?

segments $\mathbf{p}_i$ and $\mathbf{p}_{i+1}$ share three data points.

The next question to ask is whether or not this class of curves is stable in the
face of highly variable or noisy data. We have already seen that Lagrange
interpolation is extremely sensitive to noisy data. Furthermore, while Catmull-
Rom curves are smooth and often more appropriate than piecewise Lagrange
interpolation, they are nevertheless interpolating curves, which means that the
error in the data is interpolated as well. Figure 3.10 compares the result of
piecewise cubic Catmull-Rom interpolation against the corresponding B-spline
curves on the same data. Notice that the B-spline curves do not interpolate the
data, as predicted. More importantly, these curves are much more "sluggish"
than the interpolating curves: they smooth out the variations among nearby points.
A physical analogy is that they have greater inertia than the interpolating curves.
We can conclude that the B-spline curves are much less susceptible to small-scale
variation in our samples.

Let us now consider the behaviour of B-spline curves when the sample data
are taken from a discontinuous function. Figure 3.11 depicts the B-spline curves
resulting from sampling the same step function as before. Observe the complete
absence of ringing. On the other hand, the fit of these curves to the step is rather
poor. Certainly the shape of the curve or trajectory only weakly resembles the

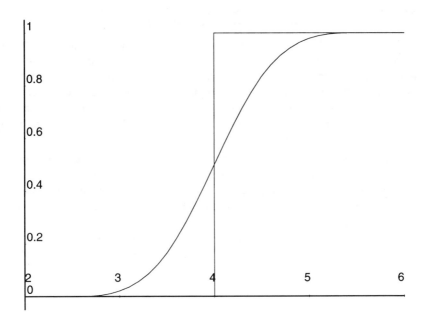

Figure 3.11. A B-spline approximation to a step function.

original step function. Depending on our application, however, this may be a satisfactory trade-off. If we are willing to sacrifice accuracy, non-interpolating piecewise polynomial approximations are a good alternative to the interpolating forms we considered earlier. In our applications of drug regulation and musical signal reconstruction, it is reasonable to say that the conservative B-spline approach is probably better, or at the very least safer, than the other approaches we have considered.

3.4. Signals and Filters[2]

A data source is *noisy* if data derived from that source have a tendency to contain errors. If we know something about the source of our data, then we can often design preprocessing approaches that enhance attractive and inhibit undesirable properties of the data. For example, medical scanning technologies such as magnetic resonance imaging (MRI) exhibit characteristic (statistical) noise patterns that left untreated would render the data meaningless. By *filtering* the

---

2. This section is notationally dense, despite the fact that the concepts are not difficult. The need for this much notation will become clearer with time.

data, however, we can extract good representative values of the quantities the device has measured. The consequence is a 3-D imaging system of remarkable resolution, and, one would hope, great diagnostic value.

The theory and processing of *signals* is of monumental importance to our day-to-day lives. We listen to radio signals and watch reconstructed television signals; video signals are encoded in analogue or digital fashion onto videotape; continuous audio signals are digitally sampled and are recorded on compact disks; signals are broadcast to and from satellites, along wires, through fibre-optic cables, and via microwaves. In most cases such signals have undergone many different changes of representation and other kinds of treatment.

Mathematically, a *signal* is a function f over some discrete or continuous domain, usually involving time or space. Pragmatically, a signal represents a varying quantity that can somehow be measured or *sampled* at discrete moments. An exact mathematical representation of the signal as a function is usually unknown. If the signal is represented as a function with *domain* $\mathbf{R}^n$ or some subset, it is called a *continuous* time or space signal in the sense that the domain for the signal is continuous set.[3] If the signal has domain $\mathbf{Z}^n$, then it is called a *discrete* time or space signal. Frequently a discrete signal has actually been derived from a continuous source. If the elements of a discrete signal represent values from a continuous signal at specific times or positions, then these elements are called *samples*. Once we have sampled a continuous signal, then what we work with afterwards is a discrete signal, or simply a sequence of values. We may wish to operate on the discrete signal, and then later turn it back into a continuous signal. Our trajectory problem fits into this category. An interesting issue that we shall consider later is that of the minimum number of discrete values that are needed to represent a continuous signal.

Exercise 3.5. Do this exercise before reading past it. Suppose we have a long sequence of our course grades and the semester during which the courses were taken. Given this list, we would like to come up with a single grade that gives a good measure of our performance. We could, for example, take the maximum, or the minimum, but most would agree that these are not fair indicators. Define several other schemes that might be better. Now suppose our list is a long, ever-growing sequence, which is not far from the truth for some of us. We would like to be able to slide our measure back and forth in time to give a sense of our score at arbitrary points in time. Do all measures work equally well at distinguishing our performance?

As we mentioned earlier, another very common problem is that of coping with data values that are "noisy," or known to be inaccurate. If we know something precise about the inaccuracies, we could design a scheme to compensate directly

---

3. Consequently, a continuous-time signal is not synonymous with the notion of a *continuous function* in calculus, since a continuous-time signal could easily be discontinuous.

for them. For example, a device might consistently report an object as being 5 cm further away than it actually is, in which case there is an obvious fix. More often, however, our knowledge of the inaccuracy is itself inaccurate. A general approach to treating this class of noise is by the use of filtering. The idea underlying the approach is very simple: rather than taking the sample value at a given instant as the only indicator of what really is going on at that time, we make the plausible assumption that recent samples (possibly in the past and the future) also provide information about that instant. In the case of our trajectory problem, suppose we have a partial sequence (or signal) **P** of values

$$\cdots, P_{i-2}, P_{i-1}, P_i, P_{i+1}, P_{i+2}, \cdots . \qquad (8)$$

We shall use boldface to represent sequences of values defining signals and filters. Specific values of a sequence will be in italic. Since we know that these samples are related to what we expect to be a continuous path of a fly, we are justified in thinking that samples close to time i are all indicators of the behaviour of the fly at time i. Rather than reporting P_i as the position of the fly at time i, we might instead form a weighted averaging of nearby samples together with P_i. On the assumption that, say, two samples before and two samples after are also relevant to the position of the fly at time i, we could treat the samples as follows:

$$\bar{P}_i = w_{i-2}P_{i-2} + w_{i-1}P_{i-1} + w_iP_i + w_{i+1}P_{i+1} + w_{i+2}P_{i+2}. \qquad (9)$$

This set of operations is depicted in Figure 3.12. Notice how this filtering operation is readily transferable to special-purpose arithmetic hardware.

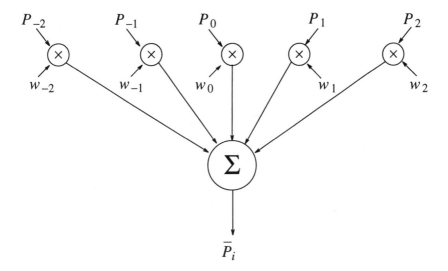

Figure 3.12. Computing a single, filtered value.

If we set $w_k = 0$ for $k \neq i$ and $w_i = 1$, then

$$\overline{P}_i = P_i, \tag{10}$$

so that this is actually a more general way of looking at the original input data. The set of weights w is called a *filter*. We shall see several examples of filters shortly. Normally, the weights are derived from a filter function f, which, for each value in the domain of the filter, gives a weighting in return. More precisely, a one-dimensional filter is a function $f : A \to \mathbf{R}$, where A is either $\mathbf{R}$ or $\mathbf{Z}$ depending on whether the filter has a continuous or discrete domain. Furthermore we shall (usually) insist that the filter is *normalised*: if f is continuous, then

$$\int_{\mathbf{R}} f = 1, \tag{11}$$

and for a set of discrete weights,

$$\sum_{i \in \mathbf{Z}} w_i = 1. \tag{12}$$

Thus f behaves like a *probability distribution*, except that there is no requirement that f (or the w) must always be positive. Typically, f is defined as a function centred at 0. The subregion S of A over which the values of f are not uniformly zero is called its *support*. Thus depending on the domain of f, S is either a subset of integers or a real interval. The support for f can be finite or infinite, though for f to be practical, it must have finite support. In fact, practical filters typically have *local* support, meaning that only a small number of nearby samples is used to define a filtered result. The filter given by the weights in Eq. 9 has local support.

Eq. 9 and the accompanying figure shows us how to compute a single, filtered value $\overline{P}_i$ given a original sequence of values $\mathbf{P}$, where

$$\mathbf{P} = \cdots, P_{-2}, P_{-1}, P_0, P_1, P_2, \cdots. \tag{13}$$

To get a new sequence of filtered values

$$\overline{\mathbf{P}} = \cdots, \overline{P}_{-2}, \overline{P}_{-1}, \overline{P}_0, \overline{P}_1, \overline{P}_2, \cdots, \tag{14}$$

we need to apply a sequence of filters (or corresponding set of weights)

$$\mathbf{F} = \cdots, \mathbf{F}_{-2}, \mathbf{F}_{-1}, \mathbf{F}_0, \mathbf{F}_1, \mathbf{F}_2, \cdots \tag{15}$$

to $\mathbf{P}$, meaning that each $\overline{P}_i$ is computed by applying filter $\mathbf{F}_i$ to $\mathbf{P}$. Note that each $\mathbf{F}_i$ is itself a sequence of values, namely a sequence of weights. Figure 3.13 illustrates a common way to think of filtering: a black box that on input $\mathbf{P}$ will produce $\overline{\mathbf{P}}$. We can therefore think of filtering as a mapping from functions to functions.[4]

---

4. This may seem like a strange concept, but notice that symbolic differentiation and indefinite integration are two more familiar examples of operations that map functions to functions.

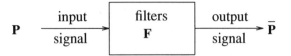

Figure 3.13. The effect of applying a sequence of filters **F** to the signal **P**.

The definition of **F** is very general: we could, for each $\bar{P}_i$ to compute, define a different filter function $\mathbf{F}_i$ that specifies the weights to be applied to **P** in order to form $\bar{P}_i$. The filter $\mathbf{F}_i$ corresponding to the weights in Eq. 9 is

$$F_i[j] = w_j, \quad j = i-2, \cdots, i+2. \tag{16}$$

Rewriting the filtering operation of Eq. 9 in a more general way,

$$\bar{P}_i = \sum_{j=i-2}^{i+2} F_i[j]P_j. \tag{17}$$

The use of square brackets indicates that the domain of the filter is integral. Of course, there is nothing magic about the *half-width* of the filter being two. For generality, we define a filter to be uniformly zero outside its support. We likewise assume the domain of the signal can be extended to all of **Z** by setting it to zero outside of its original domain. This allows us to rewrite Eq. 17 as

$$\bar{P}_i = \sum_{j=-\infty}^{+\infty} F_i[j]P_j. \tag{18}$$

A mathematical definition of an object rarely gives a good indication as to how to compute it, and this definition of *convolution* is no exception. However it allows us to talk easily about filters of arbitrary support. We shall address the issue of computation shortly. For some applications, we may actually wish to define a separate filter $\mathbf{F}_i$ for each $\bar{P}_i$ we wish to compute, but in general this is not practical. An alternative approach is to define a single filter function that is centred at zero, and to "slide" it into place by centring it at position i as needed. In Eq. 9 for example, suppose that for each $\bar{P}_i$, we wish to apply the same filter as defined in Eq. 16. Then we define a single prototype filter **f** centred at zero:

$$f[j] = w_j, \quad j = -2, \cdots, +2. \tag{19}$$

In other words, we have set i to zero. To reconstitute $\mathbf{F}_i$ for any i, we shift the weights $f[j]$ making up **f** so that the filter is centred over position i:

$$F_i[j] = f[j-i] = w_{j-i}, \quad j = i-2, \cdots, i+2. \tag{20}$$

It turns out that it is notationally more convenient to reverse the sense of i and j in the above.[5] That is, we instead shall define $F_i[j]$ as

$$F_i[j] = f[i-j] = w_{i-j}, \quad j = i-2, \cdots, i+2. \tag{21}$$

There is only one small problem with this definition: the filter values (and therefore the weights) are reflections of the weights through the $x = 0$ axis. Since most filters are even functions—a function f is *even* if $f(x) = f(-x)$—this is irrelevant. In general, however, we may have to flip our filters about the y axis.

Given all of this notation, we can now write down a general expression for discrete filtering, called the *convolution sum*:

$$\bar{P}_i = \sum_{j=-\infty}^{+\infty} F_i[j]P_j = \sum_{j=-\infty}^{+\infty} f[i-j]P_j. \tag{22}$$

When i is taken to vary over each element in the original signal **P**, then Eq. 22 is so common that it is given a name: it is the *discrete convolution* of **P** with **f**. The fact that **P** is a signal and **f** is a filter is irrelevant. All that matters is that they are both sequences of values. The notation for convolution when taken over all i is

$$\bar{\mathbf{P}} = \mathbf{P} * \mathbf{f}, \tag{23}$$

$$= \cdots, \bar{P}_{-2}, \bar{P}_{-1}, \bar{P}_0, \bar{P}_1, \bar{P}_2, \cdots. \tag{24}$$

In this case, the output signal, $\bar{\mathbf{P}}$, is the result of performing a convolution sum for each i, of the signal **P**, and a filter **f** that has been centred at i. The sequence of filters **F** in Eq. 15 is now defined implicitly using one prototype filter **f**. We are thus performing the same filtering operation regardless of our position i in the filtered signal. In this case, the filtering operation is called *space-*, *shift-* or *time-invariant*, depending on the interpretation of the signal. In some applications, it is useful to use different filters in different portions of the signal, and it is easy to see how our notation would allow such a possibility.

Convolution is commutative and associative like regular multiplication:

$$\mathbf{P} * \mathbf{F} = \mathbf{F} * \mathbf{P},$$

$$\mathbf{P} * (\mathbf{F} * \mathbf{G}) = (\mathbf{P} * \mathbf{F}) * \mathbf{G},$$

for sequences **P**, **F**, and **G**. Thus what we choose to interpret as a signal or as a

---

5. Explaining this "convenience" in greater detail would require a more rigorous treatment of this material, but it is worth providing some motivation for interested readers. In the study of signals, a *system* is a mapping from signals to signals. A large and important class of systems, *linear time-invariant (LTI)* systems, can be constructed out of some simple building blocks and operations, namely a set of possible samples, the basic arithmetic operations of summation and multiplication, and "impulse" functions to glue a sequence of samples together into a signal using a summation operation. All LTI systems are completely characterised using a convolution notation as defined in Eq. 21, as are all linear filtering operations on them.

filter is entirely up to us.

As a simple sanity check, let us define the following filter:

$$\delta[x] = \begin{cases} 1 & \text{if } x = 0. \\ 0 & \text{otherwise.} \end{cases} \tag{25}$$

This is a valid filter because the sum of the weights is one. In fact, only one weight is nonzero. This filter is called the *Kronecker* δ-function. Let us consider the computation of an arbitrary $\overline{P}_i$ using this filter. From Eq. 22,

$$\overline{P}_i = \sum_{j=-\infty}^{+\infty} \delta[i - j]P_j. \tag{26}$$

Since $\delta(x)$ is nonzero only when $x = 0$, then the only term that contributes anything to the sum arises when $i = j$. In this case,

$$\overline{P}_i = \delta[i - i]P_i$$

$$= \delta[0]P_i.$$

$$= P_i.$$

After applying this filter for each i, we get back exactly that with which we started. Compare this result with the discussion just above Eq. 10.

The notion of convolution extends to continuous signals as follows. If $f : \mathbf{R} \to \mathbf{R}$ is a (continuous) filter, and $s(t)$ is a continuous signal, then the filtered signal $\overline{s}(t)$ is given by

$$\overline{s}(t) = (s * f)(t) = \int_{x=-\infty}^{+\infty} f(t-x)\, s(x)\, dx. \tag{27}$$

While the form of convolution is different, the block diagram for continuous signals is exactly the same as for discrete signals, namely Figure 3.13. In this case, we have dropped the square bracket convention because the filter f is defined over all $x \in \mathbf{R}$. We will return to continuous-time convolution in Chapter 5, but first, a couple of remarks. First, if the filter domain is a subset of $\mathbf{R}$, then obviously the domain of integration in Eq. 27 can be tightened up accordingly. Second, the definition requires that in principle a potentially infeasible integral must be performed to compute $\overline{s}$ for a *single* value of t. What usually happens is that, with a little cleverness, the convolution of a signal with a specific filter can be simplified.

3.4.1. Sample Filters and Their Effect

Filters have many uses. They can be used to remove noise, to smooth data, to isolate specific behaviour, and to enhance certain effects. The area of digital filter design is extremely active, especially the problem of implementing filters in hardware. Let us examine some applications of filtering to noise and smoothing.

Our first filter, the Kronecker δ, was presented above and acted as an identity function. The next simplest kind of filter is to perform a *moving average*. The idea here is to let $\overline{P}_i$ be the outcome of averaging some values about P_i. For example, a three-point moving average would be:

$$\overline{P}_i = \frac{1}{3}\left(P_{i-1} + P_i + P_{i+1}\right). \tag{28}$$

This has the effect of smoothing out small deviations between points, especially if the points are generally fairly close in value. Eq. 28 has the following corresponding filter definition:

$$f_{avg}[i] = \begin{cases} \dfrac{1}{3} & \text{if } i \in \{-1, 0, 1\}. \\ \\ 0 & \text{otherwise.} \end{cases} \tag{29}$$

In general, a moving-average filter of support $A \subseteq \mathbf{Z}$ can be written as

$$f_{avg}[i] = \begin{cases} \dfrac{1}{\#A} & \text{if } i \in A. \\ \\ 0 & \text{otherwise.} \end{cases} \tag{30}$$

Here, $\#A$ denotes the cardinality of the set A (i.e., the number of elements in A). This filter, then, weights equally all sample points in its support. The wider the support, the smaller the weight attached to each point.

It is worthwhile to examine the effect of this filter on some sample signals. Figure 3.14 depicts the effect of passing the samples taken from the step function (Eq. 1) through the three-point moving-average filter in Eq. 29. The samples are taken every 0.1 units. Applying Eq. 22, the resulting signal is given by:

$$\overline{\mathbf{P}} = \mathbf{P} * \mathbf{f}_{avg}. \tag{31}$$

Remember that $\mathbf{f}_{avg}$ is a sequence of weights $f_{avg}[i]$ for all $i \in \mathbf{Z}$. Since $\mathbf{f}_{avg}$ is zero except for three points, we can write the computation of $\overline{P}_i$ for each i as

$$\overline{P}_i = \sum_{j=i-1}^{i+1} f_{avg}[i - j]P_j.$$

As we can see, the effect of the filter is rather minor. In regions where the step function is constant, the filtered result is the same as the original. The only difference between the two arises in the small window around $x = 4$ where the

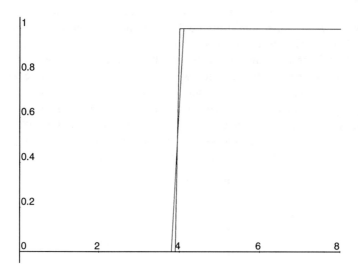

Figure 3.14. A step function and the result of passing it through a three-point moving-average filter.

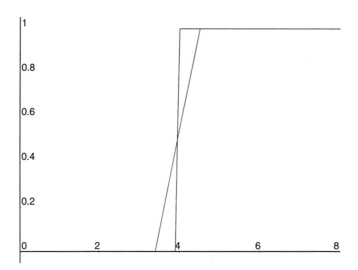

Figure 3.15. A step function and the result of passing it through a symmetric 11-point moving-average filter.

makes more of a ramp. If we let the moving average operate over a wider support with five points averaged to the left and five to the right of point P_i, for a total of 11 points, we get the result in Figure 3.15. In this case, the effect of the filter is more obvious. However, note that the filtered signal still has rather sharp corners in it. This is because the extent of the filter has a sharp cut-off. Rather than designing a more sophisticated filter, we can make the following observation. If $\overline{\mathbf{P}}$ is a smoothed version of $\mathbf{P}$, then could we compute an even smoother version by smoothing $\overline{\mathbf{P}}$? Indeed we can, and the result is called *iterated filtering*. Computer scientists might be tempted to call this *recursive filtering*, but in signal theory, this refers to a more general process. Suppose, then, we apply $\mathbf{f}_{avg}$ repeatedly to successive versions $\overline{\mathbf{P}}_1, \overline{\mathbf{P}}_2, \cdots$ of the filtered signal as follows:

$$\overline{\mathbf{P}}_0 = \mathbf{P} \tag{32}$$

$$\overline{\mathbf{P}}_1 = \overline{\mathbf{P}}_0 * \mathbf{f}_{avg},$$

$$\overline{\mathbf{P}}_2 = \overline{\mathbf{P}}_1 * \mathbf{f}_{avg},$$

and so on. In general, for $k > 0$,

$$\overline{\mathbf{P}}_k = \overline{\mathbf{P}}_{k-1} * \mathbf{f}_{avg}. \tag{33}$$

The result of iterated filtering using our 3-point moving-average filter on the step function is depicted in Figure 3.16. In this case, we are displaying $\overline{\mathbf{P}}_7$, but in fact the differences between $\overline{\mathbf{P}}_5, \overline{\mathbf{P}}_6$, and $\overline{\mathbf{P}}_7$ are minimal.

Exercise 3.6. For any sequence $\mathbf{S}$, we can write $\mathbf{S} * \mathbf{S}$ as $\mathbf{S}^2$, and more generally we can write the k-fold self-convolution of $\mathbf{S}$ as $\mathbf{S}^k$. By using the associative property of convolution, show that Eq. 33 can be written

$$\overline{\mathbf{P}}_k = \mathbf{P} * \mathbf{f}^k.$$

Figure 3.17 depicts $\mathbf{f}_{avg}^k$, for $k = 0, 1, 2, 3$. Observe that $\mathbf{f}_{avg}^1$ is a triangle. Indeed, each $\mathbf{f}_{avg}^k$ consists of $k+1$ piecewise polynomial segments of degree k. For example, $\mathbf{f}_{avg}^3$ is the cubic B-spline basis! We see, therefore, that a single filter can serve, using convolution, as the basis for a family of filters. Furthermore, an iterated filter quickly assumes the shape of *Gaussian probability distribution*:

$$g(x;\sigma) = \frac{1}{\sqrt{2\pi}\sigma} e^{-\frac{x^2}{2\sigma^2}}. \tag{34}$$

The value σ^2 is the *variance* and controls the spread of the distribution. The square root of the variance, namely σ, is called the *standard deviation*. Figure 3.18 illustrates several Gaussian distributions with different variances. The larger the variance, the flatter the function and the longer its "tail". Observe in Figure 3.19 that the filter $\mathbf{f}_{avg}^8$ with a half-width of 7 is close to a Gaussian with $\sigma \approx 11.4$. The relationship between the width of the iterated filter and σ is discussed in most probability books (normally as an outcome of the Central Limit Theorem).

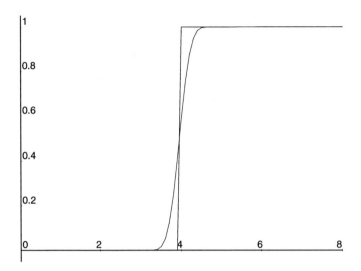

Figure 3.16. Seven iterations of a 3-point moving-average filter.

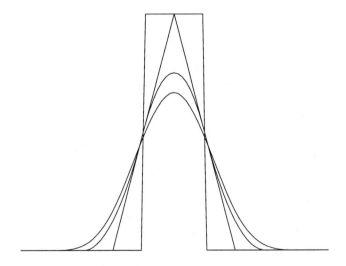

Figure 3.17. A moving-average filter and k-fold self-convolution, $k = 1, 2, 3$. Note: the filters were computed by Maple using *continuous-time* convolution.

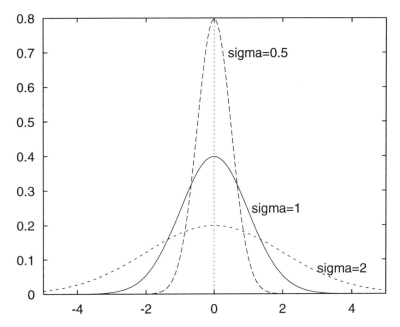

Figure 3.18. Three Gaussian distributions with variances 1, 2, and 0.5.

The support of a moving-average filter greatly affects the shape of the step function filtered using iterated filter. Qualitatively, the wider the original filter, the wider the iterated filter, which therefore blends more values together, and makes it less responsive to sharp changes in the signal. For example, the symmetric 11-point filter gives a gentler curve depicted in Figure 3.20 that looks like the B-spline approximation. On the other hand, the result of filtering with a small support moving average more closely tracks the original signal. We can thus adjust the support of the filter to tune the behaviour of the filtered signal.

The fact that the result of filtering a signal can look similar to a polynomial approximation of the data may cause one to wonder what the connection between them really is. We have already noted one connection: that an iterated moving-average filter can give a cubic B-spline. Earlier, we noted that B-splines of arbitrary degree are defined recursively. While we did not pursue this topic, it turns out that this recursion can be captured using convolution. More generally, the mathematical objects that we called *basis* or *blending* functions in previous discussions can together be thought of as filters.

Let us now consider an extremely noisy signal as found in Figure 3.21. A sine wave is plotted as well as a jittered sine wave in which a random displacement of up to ±13% was applied to each sample. Imagine that the trajectory of our well-trained fly is a sinusoid and that our bad FlyTracker is giving us this noisy data in

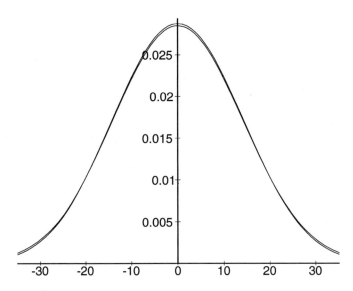

Figure 3.19. $\mathbf{f}^8$ on a moving-average filter of half-width 7 plotted together with a Gaussian with $\sigma \approx 11.4$.

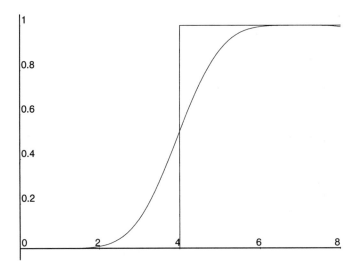

Figure 3.20. Seven iterations of an 11-point moving-average filter.

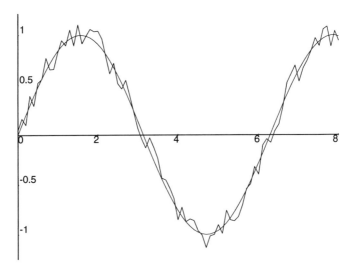

Figure 3.21. Actual flight pattern of the fly and a noisy (jittered) samples returned by tracker.

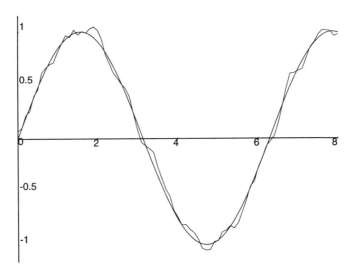

Figure 3.22. Result of filtering the noisy signal with 3-point moving average.

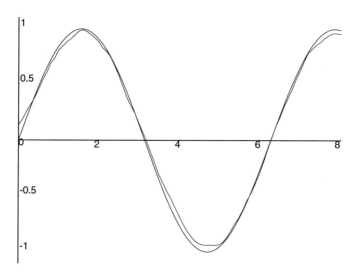

Figure 3.23. Result of filtering the noisy signal with 11-point moving average.

response. Figure 3.22 is the result of filtering the noisy signal once using the 3-point filter, while Figure 3.23 is the result of filtering using the 11-point filter. The differences are remarkable. The quality of the signal after 11-point filtering is surprisingly good, and perhaps this example begins to suggest the importance of filtering to so much of the technology we use every day. Figure 3.24 depicts the result of iterating the 3-point filter five times.

What is actually happening here? This class of filtering is called *low-pass* filtering. Mathematically, this means that the "high-frequency", or more frequently varying, components of the signal are de-emphasised, but the "low-frequency" components are kept the same, or *passed*. In fact, averaging is actually a rather poor-quality low-pass filter, but it is computationally simple, and it can be fairly effective in undemanding applications. When we discuss Fourier series, we will begin to get a better technical understanding of what this means, but a visual analogy will perhaps motivate these notions informally.

Look at this page and focus on the smallest visual detail on the page that you can. Then squint your eyes so that this detail is blurred. For some of us, just taking off our eyeglasses will suffice. What you have just "implemented" is a low-pass filter in which high-frequency information is reduced, and low-frequency information is retained.[6] What we see as blurring is analogous to the

---

6. This is a highly space-variant low-pass filter, however, since the position of your eye-lashes will affect the blurring that is going on.

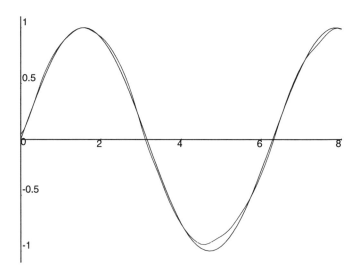

Figure 3.24. Five iterations of the 3-point moving-average filter.

smoothing effects we have seen on the noisy data. An alternative class of filters are those that are tuned to passing high frequencies and *stopping* low frequencies. Such *high-pass* filters are effective in enhancing sharp changes in a signal, and diminishing the parts of the signal that do not vary. In general, a *band-pass* filter is one in which a specific range of frequencies is passed while the others are stopped.

As may be already evident, low-pass filtering is a double-edged sword: both noise *and* highly varying detail in a signal have significant high-frequency content. Applying a badly chosen filter to a signal that contains both noise and a great deal of detail means that the detail will be thrown away with the noise. In two dimensions, for example, the step function from above would correspond to a sharp transition from a bright intensity value to a dark one (or vice versa). Trying to represent dark characters on a white background (or vice versa) is one common example of such transitions. As we can see from some of the figures above, some low-pass filters would smear this transition across many pixels, which may be undesirable, since the edges would appear overly fuzzy. In Figure 3.25, we depict the effect of three kinds of filters on a well-known image. The top left is the original mandrill image. The top right is a low-pass filtered image using a triangular filter (see below). The bottom left depicts the use of a filter that essentially takes the derivative of the image, effectively extracting thin edges. Finally, the bottom right depicts the use of an edge enhancement filter using a difference of Gaussians (see below). Both edge enhancement and edge detection are high-pass filters.

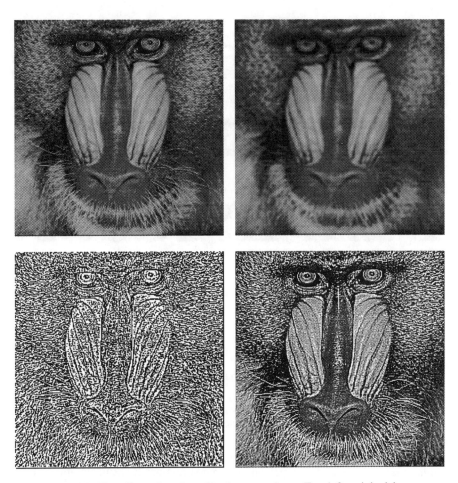

Figure 3.25. The effect of various filtering operations. Top left: original image. Top right: low-pass filtering. Bottom left: edge extraction. Bottom right: edge enhancement.

All nontrivial filtering techniques remove something from the data. For example, using a low-pass filter will cause the loss of much of the serif information in a font and the loss of crisp edges. Inevitably, *some* level of filtering is usually required for rendering high-quality characters on a computer screen, but we have to be careful about our choice of filter, lest the characters become illegible. Consider Figure 3.26. The best filter in this case would be one that trades off some visually acceptable level of blurring to cope with aliasing and reconstruction artifacts (see below) but which also preserves as much of the visual detail in the character as possible. Finding the right balance is difficult because many perceptual criteria are involved. Our understanding of the human visual system is incomplete, so these criteria are difficult to quantify.

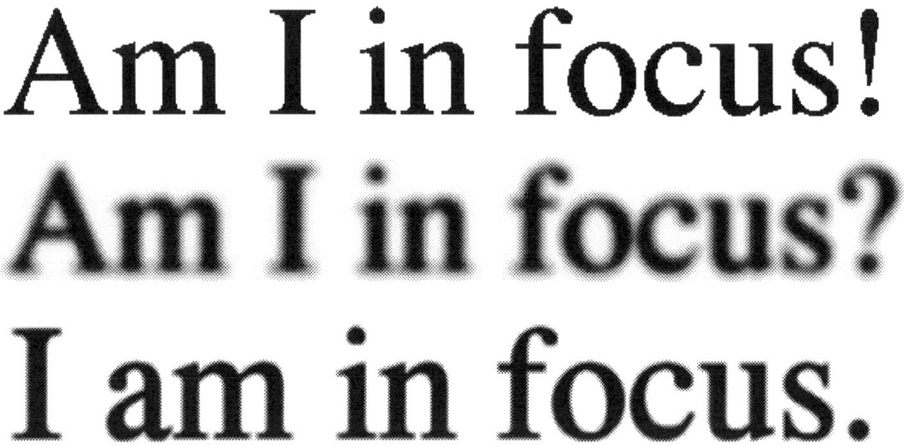

Am I in focus!

Am I in focus?

I am in focus.

Figure 3.26. Unfiltered and filtered text. The text in the top figure is unfiltered and contains noticeable "jaggies". The text in the middle figure has been aggressively low-pass filtered, resulting in a significant loss of detail. The bottom figure has been treated with a more moderate low-pass filter. A viewing distance of about 1 metre is desirable.

While the optimal rendering of characters may be difficult, one can take solace in the fact that the behaviour of the signal (i.e., the character to be rendered) is known to us. Typically, the signal is unknown and samples of it may contain significant noise. In this case, we would resort to a more aggressive low-pass filter such as a moving-average filter with wide support or a Gaussian. Note that the Gaussian not only has wide support: it has *infinite* support. In practice, *truncated Gaussian* filters, or polynomial approximations to them, are often used.

The *design* of filters to have certain behaviours with various classes of signals is an extremely interesting and rich area. We have tried to motivate some of the issues that must be faced in filter design. So far, we have only presented formally some very simple filters, those based on moving averages. We have also discussed the possibility of other filters. Our notation easily accommodates them. We shall briefly consider some slightly more interesting filters.

A *difference-of-Gaussians* filter mentioned above is often used in image processing applications. It is defined as $f(t) - g(t)$, where f and g are Gaussian distributions, with the variance of f being smaller than that of g. Recalling Figure 3.18, f would have a larger maximum than g. Figure 3.27 depicts a difference-of-Gaussians filter; the variances for f and g are one and two respectively. Notice the sharp drop-off of the filter. In regions where there is little change in the image, the filtered output gives low intensity values. Where there is an abrupt change, this filter produces large values.

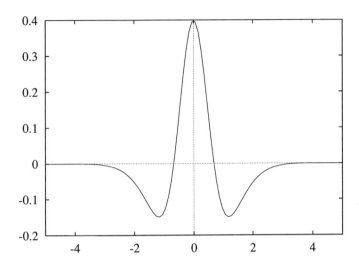

Figure 3.27. A difference of Gaussians filter.

The moving-average filter had the property that every sample point within the support of the filter was given the same weight. Consequently, this filter has a sharp cut-off, and we noticed on the step function that obvious "corners" were introduced. A *triangular* filter, which we saw earlier in the guise of an iterated moving-average filter, performs somewhat better in this respect. Such a filter of half-width w function can be defined directly as:

$$f_{tri}[i] \;=\; \frac{w - |i|}{k}, \tag{35}$$

for $i \in [-w, +w]$ (and zero outside this support). The value of k normalises the filter so that its weights sum to one. Triangular filters of half-width 3, 7, and 11 are depicted in Figure 3.28. The effect of $\mathbf{f}_{tri}$ with half-width 7 on the step signal and on the noisy sine signal can be seen in Figures 3.29 and 3.30. This filter can also be used iteratively. Figure 3.31 depicts the effect a triangular filter of half-support 3 after five iterations.

Exercise 3.7. What is k in Eq. 35 for arbitrary half-width $w \geq 1$?

Exercise 3.8. Let $k > 0$. Then $\mathbf{P} * \mathbf{f}_{tri}^{k}$ is equal to $\mathbf{P} * \mathbf{f}_{avg}^{m}$ for what value of m? (Review Exercise 3.6.)

The careful reader may have noticed an anomaly in our discussion about filtering noisy fly trajectories. Since in this case our filtering would be performed on a sequence of samples that are ordered in time, all of the filters above need to "look into the future" in order to compute a filtered value for the "present"

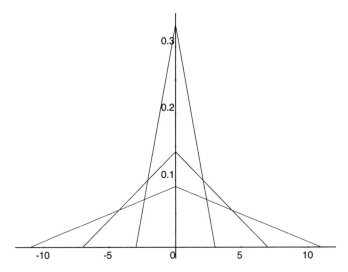

Figure 3.28. Triangular filters of half-support 3, 7, and 11.

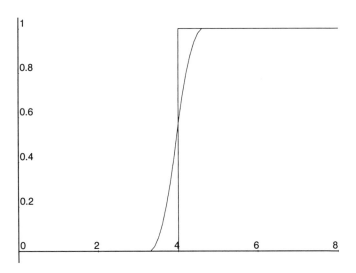

Figure 3.29. Filtered step signal with triangular filter of half-support 7.

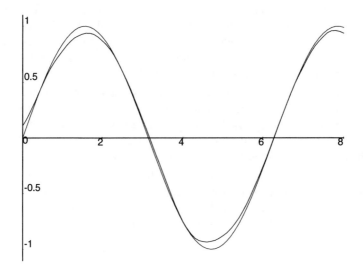

Figure 3.30. Filtered noisy sine signal with triangular filter of half-support 7.

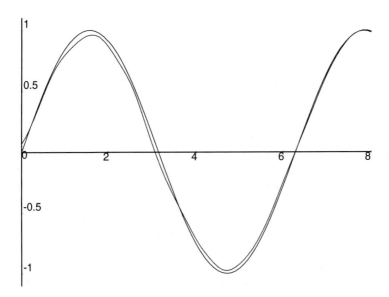

Figure 3.31. Filtered noisy sine wave after 5 iterations of triangular filter of half-support 3.

sample. With filters of wide support, this may cause a considerable lag between the filtered value and the original sequence of values. Iterated filters can also have a large lag, even for small-support filters. Filters that look into the future are called *anticipatory* filters. Filters that depend only on the current and past samples are called *causal* filters. Examples of causal filters would be the right halves of each of the triangular filters in Figure 3.28.[7]

That a plotted trajectory might lag behind our sample data is not new to us. For example, when we used Catmull-Rom interpolation, our trajectory always lagged behind by one step. In the four-point interpolant, our trajectory would lag behind by between one and four steps. Because filters may have wide support into the future, the lag time for filtered signals may be too large.

If we were to insist on completely causal filters, then we would face a difficult filter design problem, especially if our signal is very noisy. Suppose we only use the left half of triangle filter of half-support 3. Then as we see in Figure 3.32, we get very little lag (in the sense that the two signals more or less line up in time), but we get a very noisy result. On the other hand, if we use a wider extent half-triangular filter of support 11, our result is Figure 3.33. Notice here that the filtered result is less noisy, but because we are always looking backward in time, the filtered curve is slower to react to changes and consequently lags behind (i.e., is to the right of) the original signal. Another way of saying this is that the filtered result is *out of phase* relative to the original signal. Recall that the original signal is a highly jittered sine wave as in Figure 3.21; the reference sine wave allows comparison of the phases of the input and output signals.

3.5. Sampling, Filtering, and Reconstruction

The varied discussions in this chapter have prepared us for the facts that interpolation is not always successful at producing functions whose behaviour is consistent with our expectations. We have suggested the notions of approximation and of filtering to help us deal with unreliable data; indeed we have indicated that there is a close connection between filtering and polynomial interpolation and approximation. Furthermore, we have seen that even if we have at our disposal many noise-free samples of some behaviour (such as clean samples from a FlyTracker), we may not produce a good mathematical representation for that behaviour. There is a deep theory that links all of these aspects together. This is the theory of signals, but the same results in different mathematical languages can be found in other areas such as approximation theory, systems theory, Fourier analysis, functional analysis, and others. In this briefest of outlines, we shall try to put together some of these notions. The main themes are:

---

7. Recall that because of a convention used above for writing down a convolution, the "future" for a filter is to the left of the *y* axis.

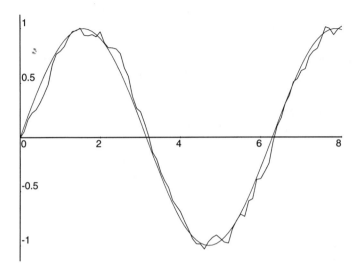

Figure 3.32. Causal triangular filter of support 3 on a noisy sine wave. The perfect sine wave is there only to indicate the phase of the original signal.

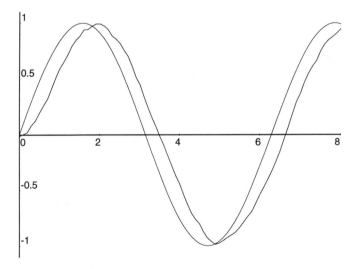

Figure 3.33. Causal triangular filter of support 11 on a noisy sine wave. Note the lag time of the filtered signal (i.e., to the right of the original signal).

- For any function we choose to represent, the *sampling theorem* provides a prescription of the minimum number of samples, or more properly the lowest sampling rate, that is required to *reconstruct* perfectly that function.

- In many cases, this minimum sampling rate is too large to be feasible. In this case, we must *filter* the signal before sampling, for otherwise we will actually be representing another signal altogether.

- Even if we can sample at or beyond the minimum sampling rate, unless we can use an ideal reconstruction (i.e., interpolation) mechanism, our resulting representation will again be incorrect.

Once again, while the theory and practice is deep, we can easily illustrate these concepts and be conscious of them when designing our fly tracking system.

3.5.1. The Sampling Theorem: An Intuitive View

How many discrete observations of a continuous phenomenon are required before we can be confident that we can faithfully represent the phenomenon? If we have some prior knowledge of the object we wish to represent, then we can design special representation schemes for the object. For example, three pieces of information are needed to represent a circle in $\mathbf{R}^2$. Four pieces of information are needed to represent a line segment in $\mathbf{R}^2$, namely two ordered pairs, or an ordered pair and a direction.

Exercise 3.9. How many different ways of specifying a rectangle can you think of?

For a polynomial of a given degree n, we have seen that there are many reasonable representations for it depending on the application, but we have seen one invariant property of the representation: $n+1$ independent observations are needed to represent uniquely a polynomial of degree n. We codified this idea by showing that polynomials can be represented as points in an $n+1$-dimensional polynomial function space. We did not, however, prove that this is the minimal dimension.

Our $n+1$ "observations" can be thought of as independent constraints on the behaviour of a polynomial. If we ask an nth-degree polynomial to satisfy more than $n+1$ constraints, then it must be the case that either one observation must after all depend on the others (i.e., it lies on the curve induced by the other points), or we have specified an inconsistent or *overdetermined* set of constraints. In this case no polynomial of degree n can satisfy the constraints, although we might produce an algorithm to find the *best* such polynomial that almost satisfies them. On the other hand, if we produce fewer than $n+1$ constraints, then our solution is *underdetermined*, meaning that a great many polynomials of degree n satisfy the constraints. Once again, we can imagine imposing other constraints such as minimal arc length that would help us to choose the most suitable polynomial.

Let us change the game slightly. Suppose we do not know beforehand the nature of the function $f(t)$ from which our samples are derived. We are in this quandary when we take samples depicting a fly's trajectory, and hope we can construct a good approximation of the continuous trajectory from a few samples. Indeed, in this case, no such function f actually exists beforehand, but it does not hurt to imagine that it does but that we just cannot get at it. The same problem arises when we digitally record a musical signal in a concert hall or in many other similar applications. We may have some qualitative information that later might help us to choose from among several representations, but let us assume that the only available quantitative information is a sequence of (noise-free) samples

$$\mathbf{P} = (t_0, f(t_0)),\ (t_1, f(t_1)),\ \cdots, (t_n, f(t_n)),\ \cdots, \tag{36}$$

for equally spaced positions (or *knots*) $t_i = i\Delta t$. An appropriate question to ask is: how many samples do we need to take before we can be confident $\mathbf{P}$ is a good representation for f over a desired domain of t? This problem is so loosely stated and so general that it seems unbelievable that it can be solved. In fact, it has been, and we shall devote the rest of this section to illustrating the basic concepts underlying the *sampling theorem*. The sampling theorem relates the ''frequency'' response or *spectrum* of a function to the sampling rate needed to represent (or encode) it. But the theorem only applies if the spectrum is bounded. The results of this theorem, and a closely related one describing the maximum quantity of information that can be transmitted along a wire of fixed bandwidth, pervades our technological achievements of the last half of the 20th Century.

We saw above in Figure 3.5 that if we do not take enough samples of a function, then our representation can be arbitrarily bad. Let us start once again from the sampling of a trigonometric function to see if we can glean the basic mechanics underlying the sampling theorem. Consider a graph of the sine function on the interval $[-4\pi, 4\pi]$ in Figure 3.34. For concreteness, assume the horizontal axis is time, and that we have a beautifully trained fly going in a strict sinusoidal pattern. The sine function has a *period* of 2π, meaning that for any integer k,

$$\sin(t + 2k\pi) = \sin t. \tag{37}$$

Exercise 3.10. What is the period of $\sin 2t$, or more generally, $\sin nt$, $n \geq 1$?

The *frequency* of the sine (and cosine) function can be written down as the number of periods or cycles it makes per unit of time. That is, the frequency of $\sin t$ is one cycle every 2π seconds, or $1/2\pi$ cycles per second (cps). We shall write the frequency of sine as

$$\omega = \frac{1}{2\pi} \text{ cps.}$$

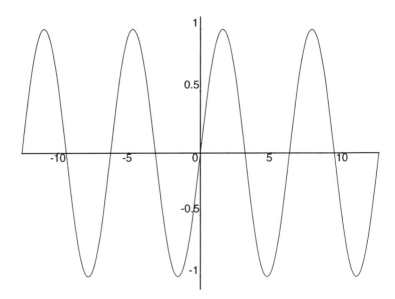

Figure 3.34. A graph of $y = \sin t$, $t \in [-4\pi, 4\pi]$.

Suppose we sample $y = \sin t$ using $k > 1$ equally spaced values of t on the interval $[-4\pi, 4\pi]$, and that we use Catmull-Rom interpolation afterwards to construct an approximation to y only from the k samples. Since we are actually interested in seeing how we can represent functions on arbitrary domains, a more appropriate way to measure the economy of our representation is not the absolute number of our samples, but rather either their spacing or their *density*. In the specific case, the spacing in t between samples is $8\pi/(k - 1)$. The sampling density or sampling *rate* is simply the reciprocal of the sample spacing. When k = 5, the sample spacing is $8\pi/4$, so the sampling rate is $4/8\pi$ samples per second. However, the value of the function at these corresponding sample points all coincide with zeroes of $\sin t$, giving us a straight line as our approximation. Increasing the number of samples over our interval to nine halves our sample spacing to $8\pi/8 = \pi$, but this still coincides with zeroes of $\sin t$, as can be seen in Figure 3.35.

When $k = 13$, the sample spacing of $3\pi/4$ is sufficient to give non-zero values for $\sin t$. The sampling rate is now $4/3\pi$ samples per second. Figure 3.36 gives the resulting curve, which is still a poor approximation.

When we increase the number of samples to 17 over the interval, giving a sample spacing of $\pi/2$ and a sampling rate of $2/\pi$ samples per second, something remarkable happens. As we can see in Figure 3.37, we now have a good approximation to the curve, and furthermore if we increase the sampling rate any further, say to 25 samples as in Figure 3.38, our approximation is not significantly improved. The threshold at which we suddenly began approximating $\sin t$ well

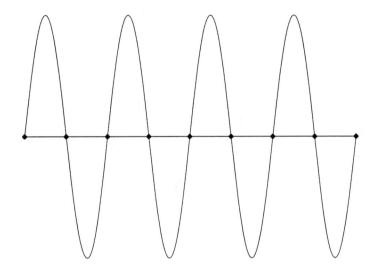

Figure 3.35. An ''undersampled'' approximation to sine using 9 samples.

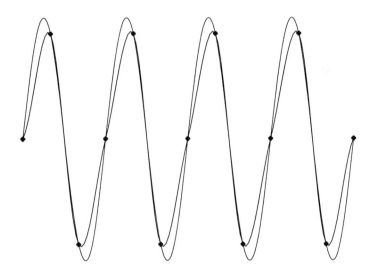

Figure 3.36. An ''undersampled'' approximation to sine using 13 samples.

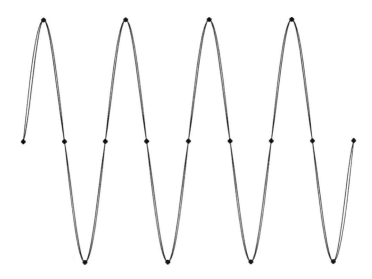

Figure 3.37. A "sufficiently" sampled approximation to sine using 17 samples.

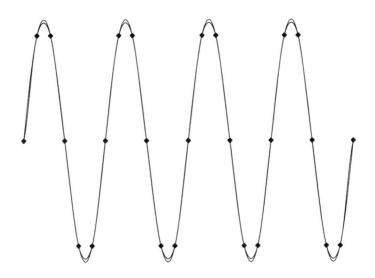

Figure 3.38. An "oversampled" approximation to sine using 25 samples.

occurred when the *sampling rate exceeded a specific multiple of the frequency of the function.* In our case, that multiple is four, since

$$\frac{16}{8\pi} = \frac{2}{\pi} = 4\omega.$$

The sampling theorem provides a lower bound for the sampling rate: it states that if a function f has a maximum frequency of ω_m (if it exists), then our sampling rate must exceed $2\omega_m$. The value $2\omega_m$ is so special it is given a name in the signal processing literature: it is the *Nyquist sampling rate*, though adding the name *Shannon* would be fair. The sampling theorem is beautiful, but as stated, it begs two questions. The first is: what is the best interpolation technique to accomplish perfect reconstruction of f? The proof of the theorem (see Chapter 5) in fact tells us that there is precisely one interpolant that is optimal, and we shall see it in the next section. The second is: what does it mean for arbitrary functions to have "frequencies"? We seem able to make sense of frequencies for sinusoids such as $\sin t$, $\cos t$, and in general for $\sin nt$, $\cos nt$, for $n \in \mathbf{Z}$. This is our clue. Suppose we can represent an arbitrary function f in terms of a summation of sines and cosines of this form. That is, suppose f can be approximated by:

$$f = \sum_{n=0}^{N} \left(a_n \cos nx + b_n \sin nx \right), \tag{38}$$

where $a_n, b_n \in \mathbf{R}$ are carefully chosen weights. Then the "maximum frequency" of f would be directly related to the largest n for which a_n or b_n is nonzero in this series representation for f. If N is the largest such n, then the highest frequency is $N/2\pi$. Observe that there is no guarantee that a finite N exists, in which case the function cannot be exactly represented using a finite sampling rate. The representation of f in terms of a summation or *series* of trigonometric functions is the *Fourier series* for f, and we shall discuss this representation later.

If a function f is uniformly sampled at a rate of at less than twice its maximum frequency, then it is necessarily *undersampled*, and there is no hope for exactly representing f. In fact, what is created is the representation for some function f' that is of necessarily *lower* maximum frequency than f. In signal processing, f' is called an *alias* for f, and this phenomenon is called *aliasing*. Because an imperfect interpolant was used in the above figures illustrating undersampling, none of the figures above technically are illustrations of real aliasing, which is why we put quotation marks around terms such as "undersampled" in the figures above. But in fact, they are quite suggestive of real thing.

If a function f is sampled at beyond twice its maximum frequency, then in principle some of the samples are redundant, and f is *oversampled*. However, for many reasons, it is usually desirable to sample a function at a significantly higher rate than the minimum. Notice, for example, that the best that we were able to do in our experiment using the sine function was *four* times its maximum frequency. We shall discuss the reasons for this shortly.

3.5.2. Reconstruction

The sampling theorem relates the minimum sampling rate of a function to its maximum "frequency." However, the other side of the coin is that unless we choose a good approach to interpolate the samples, we shall introduce substantial errors in the approximation. If we perform perfect reconstruction, then this allows us to take a continuous signal **P**, represent it in terms of a fixed set of samples, transmit it elsewhere if needed, and then reconstruct it perfectly as in Figure 3.39.

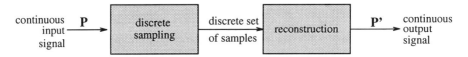

Figure 3.39. A block diagram depicting sampling and reconstruction steps.

As we have seen before, and as we see in Figure 3.40, there are many possible approaches to reconstructing continuous functions from a discrete set of samples.

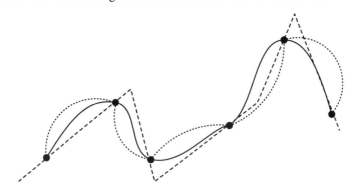

Figure 3.40. Three possible reconstructions on the same set of samples.

However, the proof of the sampling theorem shows that *only one* reconstruction is ideal (for one-dimensional signals). Suppose we have a discrete sequence of uniformly spaced samples with spacing of one unit. Then the ideal reconstruction is given by convolving with the following filter:

$$\text{sinc}(t) \quad = \quad \frac{\sin \pi t}{\pi t}. \tag{39}$$

Exercise 3.11. Show that

$$\lim_{t \to 0} \text{sinc}(t) \quad = \quad 1.$$

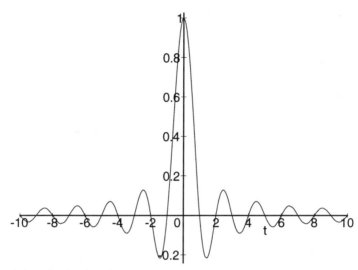

Figure 3.41. The sinc function.

A graph of the sinc function is found in Figure 3.41. Because sinc(0) is one and sinc(k) is zero for nonzero integers k, we see immediately that when convolved with a discrete signal (recall Eq. 22), on sample i the sinc function will interpolate sample P_i. On the other hand, this form of reconstruction allows one to determine the value of the output signal at non-integral point t: centre the sinc function over t and convolve it with the sequence of samples at the appropriate values of the knots. The sinc function has a significant drawback that impedes its practical use: it has a very long "tail" for both positive and negative values of t. Thus to compute a single value on the reconstructed curve involves *every* sample in the sequence of samples. This is completely infeasible as defined. It cannot be used in an online setting even if the filter is truncated, because unlike the Gaussian, which also has a long tail, the sinc has significant energy even far away from the origin. Chopping off its tail will cause ringing. We have seen, however, that piecewise interpolation has proven to be a good non-ideal reconstruction strategy. The price to pay, and which must in practice be paid, is that the sampling rate will have to be be significantly higher than twice the maximum frequency.

To summarise, ideal reconstruction filters such as sinc are almost invariably not used in practice (certainly not in the infinite-support form[8]). Cubic convolution is in fact very popular, so our use of cubic interpolation from the beginning has been a reasonable choice.

---

8. Some filters of infinite support actually become filters of finite support under certain transformations. See Chapter 5. It is also possible to reduce the ringing caused by truncation by *windowing*, which in effect applies another filter to soften edges of the truncated filter.

Exercise 3.12. Implement discrete convolution as defined in Eq. 22, and convolve the sinc function with various sequences of samples to verify how it performs interpolation.

3.5.3. Filtering

The sampling theorem tells us how densely we must sample a continuous function to be able to reconstruct that function later. It provides a clue to the number of samples we must put on a compact disk to represent a musical signal. Assuming we answer this nagging question of what the maximum frequency of an arbitrary function is, only one other major question remains: what should we do if we cannot sample as often as is as required? This amounts to saying that we are often in a position of being unable to capture all of the details present in a signal.

We know already that if we blindly undersample a signal, we will get an aliased signal as the outcome. The aliased signal is rarely a good approximation of the original signal. We cannot avoid representing a different signal if we use too few samples, but at least we can try to control the kind of signal that we get after reconstruction. The way to do this is to change the game: filter the original signal *prior* to sampling to remove the offending high frequencies. This requires convolving the original signal (perhaps performing a continuous convolution rather than a discrete one) with a *low-pass filter*. We saw several examples of such filters earlier.

Having already visited the topic of filtering, the option of low-pass filtering is clear to us. That it is the best course of action is less clear, but there is some strong evidence that it is appropriate. First, low-pass filtering amounts to softening edges in images, and reducing the sharpness of transitions in one-dimensional signals, leaving the other parts of the signal with little change. This seems to be desirable. Second, noise can manifest itself at all frequencies, but is particularly good at making high-frequency information less perceivable. Thus passing a noisy signal through a low-pass filter would have the doubly desirable effect of reducing the high-frequency content of the signal arising both from noise and from unrepresentable detail. Third, special filters can be tailored to the behaviour of certain classes of signals. For example, filters in digital audio applications can be "tuned" to the particular kinds of transient (or variable) behaviour that arises in musical signals rather than signals from our favourite extraterrestial beings or other such signals. Our chain of operations now looks like Figure 3.42.

A good graphical illustration of the use of filtering can be found in Figure 3.43. The spheres are *texture-mapped* using a rectangular texture composed of vertical stripes. The vertical stripes are mapped onto a sphere via its parameterisation. Since a sphere can be parameterised over a rectangular grid (recall Chapter 1), it is easy to superimpose a rectangular texture over the rectangular domain of the sphere's parameterisation. When we look at the object, however, we get a significant amount of "pinching in" of the texture at the poles

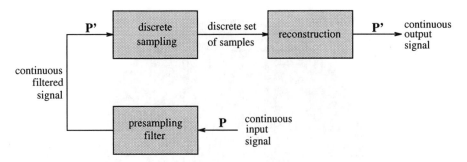

Figure 3.42. A block diagram depicting filtering, sampling, and reconstruction.

of the sphere, as we would expect. Thus an arbitrarily high amount of detail is compressed into the region near the poles, resulting in an extremely high-frequency signal that is a worst-case test of any filtering algorithm. If we sample the textured sphere at a fixed sampling rate without filtering, we get the upper image. Note the odd patterns throughout the image. These so-called *Moiré patterns* are a strong visual example of aliasing at work. If we are careful to employ filtering, however, we get the lower image. The filter used is a two-dimensional truncated Gaussian distribution. The filtered image is ''less sharp'' than the unfiltered image, but it is of more pleasing visual quality. Some aliasing problems are still evident, but the most glaring problems have been avoided.

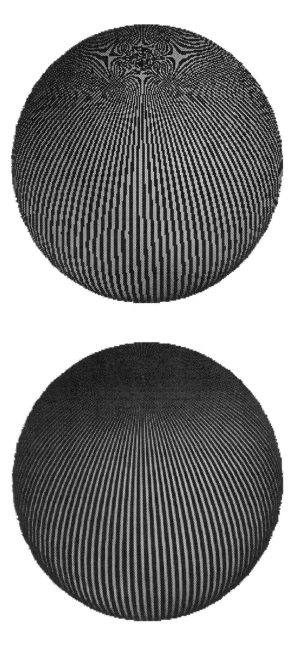

Figure 3.43. Above: Unfiltered texture mapped sphere. Below: texture mapped sphere filtered with Gaussian filter. [Image courtesy of Robert Lansdale, Dynamic Graphics Project, University of Toronto.]

Supplementary Exercises

Exercise 3.13. If a filter f has a support σ, what is the support of f^k for arbitrary $k \geq 0$?

Exercise 3.14. In a derivation following Eq. 26, we saw that for an arbitrary signal s, and the filter f given by $f[i] = \delta[i]$, where δ is the Kronecker delta function,

$$s * f = s.$$

Now, let f instead be such that $f[i] = \delta[i-k]$ for an arbitrary but fixed $k \in \mathbf{Z}$. What is $s * f$?

Exercise 3.15. Write a Maple procedure **filter(s,f)** that applies discrete filter f to discrete signal s. We'll use bold font to denote signal and filter s and f, and we'll use the typewriter font to denote their representation in Maple. Assume s denotes the values

$$s = s_1, s_2, s_3, \cdots, s_n,$$

and f is

$$f = f_L, f_{L+1}, \cdots, f_{-1}, f_0, f_1, \cdots, f_{R-1}, f_R,$$

where $L < 0$ is the support of the filter to the left (i.e., into the future) and $R > 0$ is the support of the filter to the right (i.e., into the past). Assume that s_i is zero for any $i > n$ or $i \leq 0$. To compute a filtered value $\bar{s}_i$, perform convolution as in Eq. 22. Use the assumption about s to handle cases at either end of s when you run out of samples to filter. Both s and f are to be represented as lists of the form **s = [s1, s2, ..., sn]** and **f = [f1, f2, ..., fm]**. Thus the elements of the signal s are **s[k]**, starting from **k=1**, and likewise for the filter f. This also means you will have a slight complication to index the elements of f. Assume that **f[1]** is f_L, that **f[nops(f)]** is f_R, and that **f[trunc(nops(f)/2)+1]** is f_0. So, for example, if f contains five values, then **f[1]** denotes f_{-2}, **f[3]** denotes f_0, **f[5]** denotes f_2, meaning that $L=-2$ and that $R=2$. Test your Maple code by plotting the effect of filtering signals with various filters.

Exercise 3.16. Once your filtering code is working, use it to derive a *digital reverberation* filter. Applying this filter to a signal results in making a certain number of copies of the signal, displacing them in time, and scaling (i.e., reducing the volume of) each copy of the signal. If the signal were interpreted as a sound signal, then the effect would be an echo. Write a Maple procedure that returns a filter suitable for use by the **filter** procedure above. The form of the reverb filter should be

reverb (n, dt, decay)

where **n** is the number of times the signal should be reproduced, **dt** is the spacing between the *onset* of each copy of the signal, and **decay** $\in [0,1]$ specifies how much the "volume" of each copy of the signal should be reduced: copy i of the signal should be proportional to **decay**^{i} of the original signal's value, with the constant of proportionality being such that the filter normalises to the value of one. Otherwise it would be an *amplifier* as well as a filter (which actually is common in some applications). The first copy of the echoed signal should coincide in time with the onset of the original signal, with the specified drop-off and delay from that point onward.

By way of example, a "box" function is a reasonable model of a drum beat. If we define a box function in Maple as

```
box  := [0$20, 1$10, 0$50];
```

then the result of

```
filter(box, reverb(3,15,0.3))
```

should look like Figure 3.44. The tall bump is the original box signal, while the train of small bumps is the filtered (reverberating) signal. Notice that the onset of the first copy coincides with the onset of the original signal.

This question is easier than it seems, though there are some small (non-Maple) technical hurdles to overcome. Hint: use Exercise 3.14. You may find that the "echo" occurs in the reverse order. It is also a bit tricky to make the echo properly coincide with the original signal.

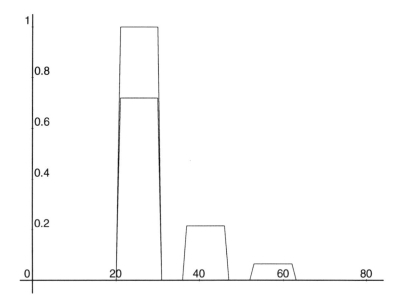

Figure 3.44. Effect of "echo" filter.

Chapter 4
Computational Integration

Integration is one of the most common of scientific computations. Sometimes a symbolic mathematical system or sheer ingenuity allows us to derive a closed-form solution (i.e., an algebraic equation) for an integral. However, the integrals of many simple functions do not have simple solutions, and consequently much effort has been devoted to performing numerical definite integration robustly and quickly. In this chapter we shall take a brief tour through the great variety of integration techniques, and we shall explore the interpolation mechanisms we developed earlier as possible bases for good numerical integration routines. In particular, we shall discuss Riemann sums, trapezoid rule, Simpson's rule (and other approaches based on Lagrange interpolation), and smooth cubic interpolation. These correspond to polynomial interpolation of degrees zero to three. We then show how to extend the techniques to *compound* or piecewise integration. Several case studies will be considered, but an error analysis will be deferred to the next chapter. Finally, we shall briefly consider the computation of integrals using a completely different approach: flipping coins. These are called *Monte Carlo* methods. We used the phrase *computational integration* for the chapter title rather than *numerical integration* or *quadrature* so as to include the symbolic computation of integrals. While it is certain that traditional numerical integration techniques will remain unsurpassed for efficiency and robustness, the ease with which symbolic systems can be used to reason about and to explore properties of integrals makes them important partners in the effort to solve integration problems.

4.1. Introduction

Integration is an indispensable tool for describing the world around us. To define the amount of light striking a surface, the length of a trajectory, the area or volume of an object, the work done over time, the energy absorbed by a medium, the distance travelled by an object, the probability that some event occurs over some time period, or a continuous-time convolution, all require integration to give them precise meaning. We have come to write the integral of a function $f: \mathbf{R}^n \to \mathbf{R}$ as

$$\int f.$$

We often wish to be more precise about the variables with respect to which we are integrating. If $f: \mathbf{R} \to \mathbf{R}$, then we may write something like

$$\int f(x) \, dx$$

for a parameter x. If f is multivariate, that is, a function of the form $f: \mathbf{R}^n \to \mathbf{R}$, then we can write

$$\int f(\mathbf{x}) \, d\mathbf{x}$$

where $\mathbf{x}$ is a point or vector in $\mathbf{R}^n$. We can be even more general, and allow f to be vector-valued, meaning that f is of the form $f: \mathbf{R}^n \to \mathbf{R}^m$. The integral for all such functions can be made precise. For this chapter, we shall focus on univariate functions of the form $f: \mathbf{R} \to \mathbf{R}$.

When the domain over which the integral is taken is not given, as is the case for those above, then such an integral is called *indefinite*. If, on the other hand, a specific finite domain is given, then the integral is called *definite*. For example, if $v(t)$ describes the velocity (in one dimension) of a particle over time, then

$$\int_a^b v(t) \, dt$$

defines the distance travelled over the definite time interval $[a,b]$. To define the distance travelled from time 0 up to any time $x > 0$, we can write

$$d(x) \;=\; \int_0^x v(t) \, dt.$$

This expression is called an *integral equation*.

Exercise 4.1. What is the indefinite integral of a polynomial in variable t of the form

$$p(t) \;=\; \sum_{i=0}^{n} a_i \, t^i \,?$$

If we allow the interval $[a,b]$ to grow by letting a approach $-\infty$ or b approach $+\infty$ (or both), then we can make sense of *improper* integrals of the form

$$\int_{-\infty}^{+\infty} f.$$

If a function f is continuous, then the *fundamental theorem of calculus* states that f has an (indefinite) integral F such that

$$\int_a^b f = F(b) - F(a).$$

Sometimes F is called an *antiderivative*, in that $dF/dx = f$.

Example 4.1. Consider the function $f(x) = \sin x$. Then

$$\int_0^\pi \sin x = 2$$

since $F(x) = -\cos x$ is such that $d(-\cos x)/dx = \sin x$. Therefore

$$\int_0^\pi \sin x = -\cos \pi + \cos 0 = 1 + 1 = 2.$$

Example 4.2. The *sinc* function was introduced in the last chapter, and has the definition

$$\mathrm{sinc}(x) = \frac{\sin x}{x}.$$

Sometimes x is replaced by πt so that the zeros of sinc πt occur when t is integral. Since the sinc function can be used as a (reconstruction) filter, it is important to be able to normalise it so that the area under the curve is one. Furthermore, it is of infinite extent in both directions. Thus we wish to know the value of the improper integral

$$\int_{-\infty}^{+\infty} \frac{\sin x}{x}\, dx.$$

Establishing this value is difficult using only real-valued calculus (also called real analysis). However, it may come as a surprise to the reader that using complex analysis it is straightforward to show that the value of this integral is exactly π.

The theory of integration has been made entirely precise using two powerful analytic approaches: Riemann sums and Lebesgue measure. The first is based on the simple notion that the integral of a function can be approximated by covering the region under the function by objects whose area we can easily compute. Rectangles are the simplest such objects, and the area under a function can be approximated by a *Riemann sum*, namely the sum of the areas of a set of very thin rectangles. As the number of rectangles increases and as they get thinner, the approximation to the integral is likely to improve. If it can be shown that the limit to this growing sequence of ever thinner rectangles converges to something, then the result is called the *Riemann integral* of the function. As we shall see, the theory of Riemann integration underlies most of our numerical integration techniques, and it is the way most of us intuitively understand the notion of integration. The theory of Lebesgue is beautiful and difficult; it generalises the notion of Riemann sums and permits the integration of troublesome functions that are not Riemann integrable. It also provides an elegant link between probability theory (in both discrete and continuous forms) and real analysis. Understanding this approach to the integral requires redeveloping most of our notions of limits, convergent sequences, leading ultimately to a more general theory of the integral. We shall happily defer discussion of this advanced topic to a future course.

Many classes of functions have closed-form indefinite integrals, and many books have been published containing both long tables and effective strategies for deriving analytic solutions to integrals. In some sense, a symbolic algebra system acts as a query language into the "database" of closed-form solutions to integrals.

Example 4.3. Maple easily finds a closed form for the integral of Runge's function:

```
> R := 1/(1+25*x^2);
```

$$R := \frac{1}{1 + 25 \; x^2}$$

```
> int(R,x=-1..1);
```

$$2/5 \; \text{arctan}(5)$$

```
> evalf(");
```

$$.5493603068$$

However, Maple is unable to find the closed-form solution for

$$\int \frac{\cos x}{1 + 25x^2} \, dx,$$

despite the fact that this function looks very similar to, and is bounded above by, Runge's function on $[-1,1]$ (see Figure 4.1). The author was not able to find a closed-form solution either. Maple reports that its definite integral on $[-1,1]$ is roughly 0.5213413.

Although many techniques that have been developed to solve integrals, including integration by parts, change of variables, series representations, limit arguments, and so on, we have conspired to devise many more classes of functions that resist symbolic or closed-form integration.

In this chapter, we shall briefly consider four approaches to integrating a univariate function: Riemann sums, the trapezoid rule, Simpson's rule and smooth piecewise polynomial interpolation. We shall see that these approaches can be very effective for a wide variety of functions.

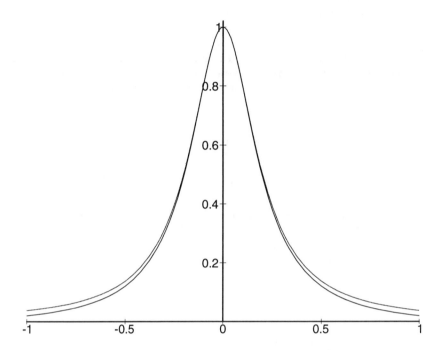

Figure 4.1. Two functions defined over $[-1,1]$: the function above is Runge's function $R(x)$, which has a closed-form integral. The lower function is simply $\cos t\, R(t)$, which Maple cannot symbolically integrate.

When dealing with integrals over a high-dimensional domain, our familiar one-dimensional integration techniques can break down. In this case, an amazingly simple but time-consuming technique based on random numbers is often used. We conclude this chapter with a discussion of *Monte Carlo* methods.

Example 4.4. Let $\mathbf{x}(t) = (x(t), y(t), z(t))$ be a parametric space curve. How long is the trajectory made by $\mathbf{x}$ on a parametric interval $[a,b]$? In other words, what is the distance along the curve between $\mathbf{x}(a)$ and $\mathbf{x}(b)$? The answer is related to the velocity of the curve. Let $\mathbf{x}'(t) = (x'(t), y'(t), z'(t))$ be the velocity vector of the curve at time t. This corresponds to the componentwise derivative of $\mathbf{x}$ with respect to t. The function $\mathbf{x}'$ defines the instantaneous *direction* of travel and the magnitude of this vector,

$$\| \mathbf{x}'(t) \| = \sqrt{x'(t)^2 + y'(t)^2 + z'(t)^2} \,,$$

defines its *speed*. The *arc length* of the curve $\mathbf{x}$ between points $\mathbf{x}(a)$ and $\mathbf{x}(b)$ is defined as

$$\int_a^b \| \mathbf{x}'(t) \| \, dt.$$

If the function is in explicit form, namely of the form

$$y = f(x),$$

then we can derive a special case of the general parametric form for the arc length. Let $x = t$, and suppose $a \leq t \leq b$. Since x varies exactly in proportion with t, $x'(t) = 1$. Furthermore, $y(t)$ is simply $f(t)$. Therefore the formula for arc length for an explicitly defined curve of the form $y = f(x)$ is

$$\int_a^b \sqrt{1 + (f'(x))^2}\, dx.$$

Exercise 4.2. Use Maple to derive the arc length of arbitrary cubic curve segments $\mathbf{x}(t) = (x(t), y(t), z(t))$ over the interval $[0,1]$.

Exercise 4.3. Use the parametric definition of a circle to show that the circumference (i.e., the arc length) of a circle of radius r has length $2\pi r$. Can this be easily extended to the arc length for ellipses? A class of functions called the *elliptic* integrals arises from the latter, so perhaps this will help to answer the question.

Example 4.5. As we indicated earlier, some very simple functions do not have closed-form solutions. Let us compute the arc length of the curve

$$y = \sin x$$

over $x \in [0, \pi]$. The arc length is simply

$$L = \int_0^\pi \sqrt{1 + (y')^2}\, dx$$

$$= \int_0^\pi \sqrt{1 + \cos^2 x}\, dx.$$

If we ask Maple to compute this integral, we find (after a long computation) that it cannot. This is justifiable, since this simple integral has no closed form solution. It can be easily computed numerically however, and we shall do so below.

4.2. Basic Numerical Quadrature

The numerical computation of definite integrals is called *quadrature* by numerical analysts. A rich variety of numerical integration techniques has been proposed. We shall consider several techniques that build on our use of interpolation in previous chapters. First, we shall consider the use of Riemann sums, which illustrates the original thinking behind the definition of the integral. Subsequently, we will consider three techniques based directly on our discussion of piecewise interpolation from Chapter 2: piecewise linear, Lagrange, and cubic Catmull-Rom interpolation. Each technique yields a way of approximating the area under a curve by computing (exactly) the areas of piecewise curves with each curve defined over a specific subinterval or *panel*. These subintervals are a partition of the overall interval $[a,b]$ over which we wish to compute the integral. As we shall see, each approximation technique will yield a different *quadrature rule*, which can be thought of as a set of function values, together with a set of weights associated with each value.

4.2.1. Riemann Sums

Suppose we wish to integrate a function such as

$$f(x) = \sin x$$

over $[a,b] = [0,\pi]$. We of course write this as

$$\int_0^\pi \sin x \, dx,$$

and from an example above, we know that the real value of this integral is 2. A graph of this function can be found in Figure 4.2.

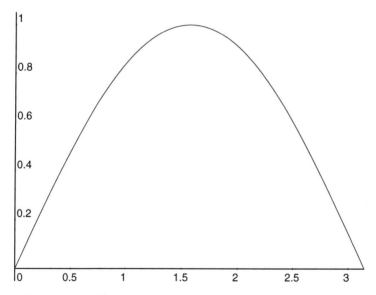

Figure 4.2.   A graph of the function $y = \sin x$ over $[0,\pi]$.

Let us pretend that this integral is difficult to solve and consider a simple method of computing it numerically: cover the region to be integrated by rectangles whose area is simple to compute; we simultaneously add more rectangles and reduce their width so that hopefully we come up with a better approximation. There are ways of constructing sequences of approximating

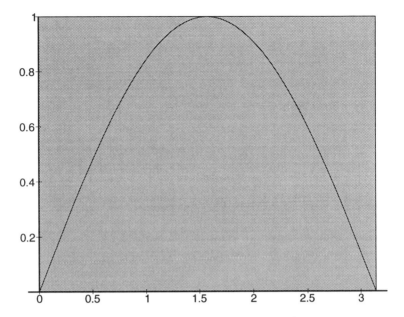

Figure 4.3.   Enclosing $y = \sin x$ over $[0,\pi]$ by a single rectangle.

rectangles. Let us start with one rectangle of width $b - a = \pi$ and of height

$$h_0 \;=\; \sin\left(\frac{b+a}{2}\right) \;=\; \sin\left(\frac{\pi}{2}\right) \;=\; 1.$$

This approximation to the integral can be seen in Figure 4.3. The area under the rectangle is simply its width times its height, which is $\pi \times 1 = \pi$. This is clearly not a good estimate, so we can split our interval $[0,\pi]$ into two intervals $[0,\pi/2]$ and $[\pi/2,\pi]$, and compute two new rectangles. One of the rectangles will have height

$$h_{left} \;=\; \sin\left(\frac{\pi/2+0}{2}\right) \;=\; \sin\left(\frac{\pi}{4}\right) \;=\; \frac{\sqrt{2}}{2}.$$

As it happens, the other rectangle will have the same height, since

$$h_{right} \;=\; \sin\left(\frac{\pi + \pi/2}{2}\right) \;=\; \sin\left(\frac{3\pi}{4}\right) \;=\; \frac{\sqrt{2}}{2}.$$

Both rectangles are of width $\pi/2$, and therefore the area of the two rectangles is

$$\frac{\pi}{2} h_{left} \;+\; \frac{\pi}{2} h_{right} \;=\; \frac{\pi\sqrt{2}}{2} \;\approx\; 2.22144.$$

The result is depicted in Figure 4.4. We may recursively proceed in this fashion until we are satisfied that we have converged to a good result. We shall discuss the issue of convergence shortly. In general, our original interval $[a,b]$ may be

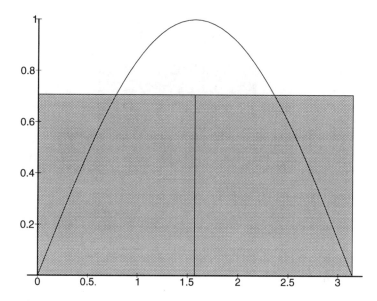

Figure 4.4. Approximating the area under $y = \sin x$ over $[0,\pi]$ by the area of two rectangles.

broken up into many subintervals with x components $a = x_0, x_1, \cdots, x_n = b$. We form n rectangles $R_0, \cdots, R_{n-1}$. Each rectangle R_i has midpoint

$$m_i = \frac{x_{i+1} + x_i}{2},$$

height $\sin m_i$, and width $x_{i+1} - x_i$. Thus the area of all the rectangles is

$$A_n = \sum_{i=0}^{n-1} (x_{i+1} - x_i) \sin m_i.$$

For a general function f, the approximation of the definite integral over n subintervals is given by the following rule.

Riemann Sums Quadrature Rule.

For each subinterval $[x_i, x_{i+1}]$, compute a single sample value $f(m_i)$, where $m_i = (x_{i+1} + x_i)/2$. If there are n subintervals, then the overall area is

$$A_n = \sum_{i=0}^{n-1} (x_{i+1} - x_i) f(m_i).$$

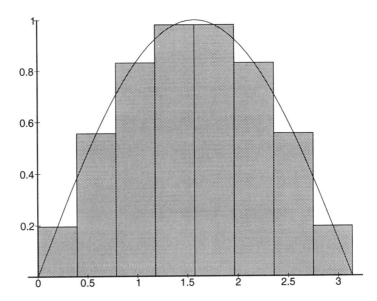

Figure 4.5.   Approximating the area under $y = \sin x$ by the area of eight rectangles.

Figure 4.5 depicts the result of using eight rectangles to approximate the area under the curve.   Under this approximation,

$$\int_0^\pi \sin x \, dx \approx A_8 \approx 2.0129.$$

Table 4.1 illustrates the effect of the number of rectangles (and their size) on our integral.   While the results in this table are for k rectangles of uniform widths, with k a power of two, our formulae above are consistent with using an arbitrary number of rectangles with nonuniform widths.   We shall have more to say about this approach later when we discuss an overall comparison, but clearly it is hard to recommend this approach when such a simple function to integrate requires about 512 samples to achieve five decimal places of accuracy.   This is a particularly serious problem if the cost of invoking the function to get samples is itself high. In such cases it is very important to get the most out of our samples.

Example 4.6.   Let us return to Example 4.5 and attempt to approximate the arc length of the function $y = \sin x$.   Figure 4.6 depicts the function and its covering by eight rectangles. Using only 32 rectangles, we find that

$$\int_0^\pi \sqrt{1 + \cos^2 x} \, dx \approx 3.820197792,$$

which is correct to eight decimal places.

n	Area (A_n)
1	3.14159
2	2.22144
4	2.05234
8	2.01291
16	2.00321
32	2.00080
64	2.00020
128	2.00005
256	2.00001
512	2.00000

Table 4.1.   The effect of the number of rectangles (and proportionally thinner widths) on the accuracy of the definite integral.

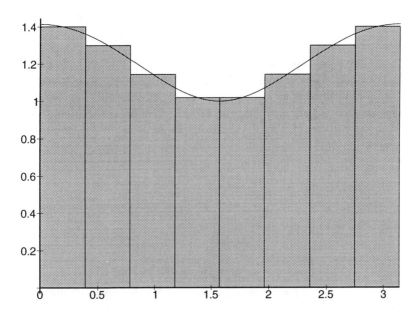

Figure 4.6.   Approximating the arc length of $y = \sin x$ by the area of 8 rectangles.

4.2.2. Integration Based on Piecewise Polynomial Interpolation

The mathematical machinery we developed in Chapter 2 has nicely equipped us to solve integration problems. In particular, we can employ piecewise polynomial interpolation using samples from the function to be integrated. Since polynomials are easy to integrate both numerically and symbolically, effective quadrature schemes can be developed. We shall first present piecewise linear, quadratic and smooth cubic approaches by considering the each curve segment individually. We then present more general formulae for *compounding* the integral over the entire domain of integration, followed by an empirical study of the algorithms. In the next chapter, we consider their accuracy more mathematically.

4.2.2.1. Trapezoid Rule

A degenerate form of piecewise polynomial interpolation occurs when the degree of the polynomial is zero. This exactly corresponds to the approach using Riemann sums discussed in the previous section. When the degree of the polynomials is one, we of course get piecewise linear interpolation, which will begin our discussion in this section.

We once again break our interval $[a,b]$ up into pieces with x components $a = x_0, x_1, \cdots, x_n = b$. We shall think of our integration problem as something like our trajectory interpolation problem. In this case, we have a set of samples

$$(x_0, f(x_0)), (x_1, f(x_1)), \cdots (x_n, f(x_n)),$$

with which to work. We can join adjacent points by a line segment, forming trapezoids rather than rectangles. The base of the trapezoid is as it was for the rectangles, but rather than forming the top of a rectangle by computing the function at the midpoint of the interval, we instead connect $(x_i, f(x_i))$ and $(x_{i+1}, f(x_{i+1}))$ together. Figure 4.7 illustrates the construction for two trapezoids. The area of a given trapezoid T_i is

$$A(T_i) = \frac{f(x_{i+1}) + f(x_i)}{2} (x_{i+1} - x_i).$$

By summing the area of the trapezoids, we have another approximation scheme, given by the following summary.

Trapezoid Quadrature Rule.
  Over each subinterval $[x_i, x_{i+1}]$, evaluate $f(x_i)$, and $f(x_{i+1})$. Each point has weight $1/2$ in computing the area under each subinterval. If there are n subintervals, the overall area is

$$A_n = \sum_{i=0}^{n-1} A(T_i) = \sum_{i=0}^{n-1} \frac{f(x_{i+1}) + f(x_i)}{2} (x_{i+1} - x_i).$$

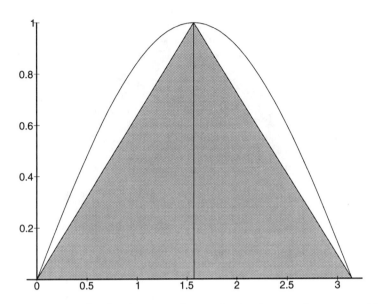

Figure 4.7.   Approximating the area under $y = \sin x$ by the area of two trapezoids
(in this case degenerate trapezoids, namely triangles).

Figures 4.8 and 4.9 depict the effect of increasing the number of trapezoids. Notice how the trapezoids more clearly track the shape of the function. Perhaps surprisingly, as can be seen in Table 4.2, the resulting approximation for a smooth function such as the sine is not significantly better than for Riemann sums. This can be explained by the fact that for fairly smooth functions, the difference in area between very thin rectangles and trapezoids is very small. We shall prove this to be the case in the next chapter.

Example 4.7.   Let us approximate the arc length of the function $y = \sin x$ using the trapezoid rule. Figure 4.10 depicts the function and its covering by eight trapezoids. Using ten trapezoids the result is again correct to eight decimal places:

$$\int_0^\pi \sqrt{1 + \cos^2 x}\, dx \;=\; 3.820197718.$$

4.2.2.2. Simpson's Rule: Piecewise Quadratic Lagrange Integration

There is nothing to stop us from employing higher-degree polynomial interpolation. One possibility would be to apply piecewise polynomial Lagrange interpolation. We first consider piecewise quadratic interpolation.

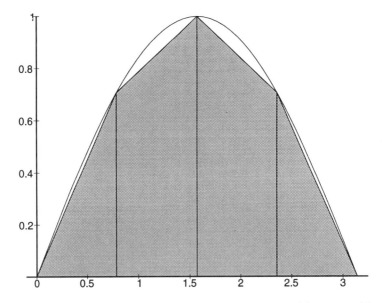

Figure 4.8.   Approximating the area under $y = \sin x$ by the area of four trapezoids.

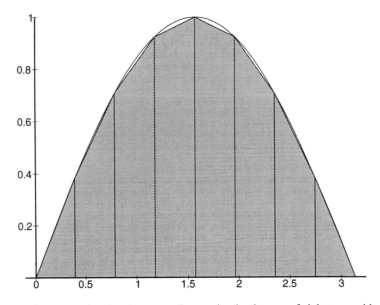

Figure 4.9.   Approximating the area under $y = \sin x$ by the area of eight trapezoids.

n	Area (A_n)
2	1.57079
4	1.89611
8	1.97423
16	1.99357
32	1.99839
64	1.99959
128	1.99989
256	1.99997
512	2.00000

Table 4.2. The effect of the number of trapezoids (and proportionally thin widths) on the accuracy of the definite integral of $\sin x$.

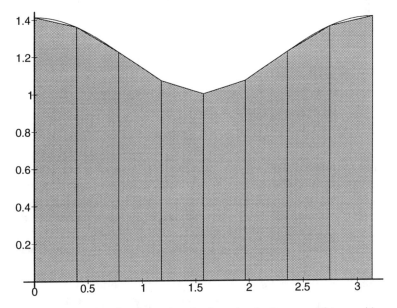

Figure 4.10. Approximating the arc length of $y = \sin x$ by the area of 8 trapezoids.

Suppose for each interval $[x_i, x_{i+1}]$ we derive the quadratic polynomial Q_i passing through the points

$$f(x_i), \quad f(\frac{x_i + x_{i+1}}{2}), \quad f(x_{i+1}).$$

This is just three-point Lagrange interpolation, and can be easily derived (in several ways) from the work in Chapter 2. We could then integrate each Q_i over $[x_i, x_{i+1}]$ and sum the result. This technique is called *Simpson's rule*.

It is instructive to use Maple to derive this result. Suppose we have a Maple routine called **ApproxLagrange(f, [a,b], n, x)** which returns the Lagrange polynomial in variable x that interpolates the value of f for $n + 1$ equally spaced points over $[a,b]$. The degree of the polynomial is therefore n. Thus in our case we wish to invoke

$$Q := \textbf{ApproxLagrange(f, [a,b], 2, x)}.$$

After simplifying, sorting, and collecting terms we still have a messy expression:

$$\frac{(2x^2 - xa - 3bx + ba + b^2)f(a)}{(a - b)^2} + \frac{(2x^2 - 3xa + a^2 - bx + ba)f(b)}{(a - b)^2} +$$

$$\frac{(4xa - 4x^2 + 4bx - 4ba)f(\frac{a}{2} + \frac{b}{2})}{(a - b)^2}.$$

This can be symbolically integrated over $[a,b]$ to yield our approximation. After asking Maple to integrate, collect and simplify, we get

$$-\frac{af(a)}{6} - \frac{af(b)}{6} - \frac{2af(\frac{a}{2} + \frac{b}{2})}{3} + \frac{bf(a)}{6} + \frac{bf(b)}{6} + \frac{2bf(\frac{a}{2} + \frac{b}{2})}{3}.$$

After some human fine-tuning, we have the following rule.

$$\int_a^b Q(x)\,dx = \frac{b - a}{6}f(a) + \frac{b - a}{6}f(b) + \frac{2(b - a)}{3}f\left(\frac{a + b}{2}\right).$$

As with the other techniques, we can divide an overall interval $[a,b]$ into n subintervals. If our aim is to compute a definite integral, then nowhere do we actually compute the quadratic curves for each subinterval, although Appendix A shows how to compute these interpolating polynomials explicitly. For each subinterval $[x_i, x_{i+1}]$, we apply the above formula by setting $a = x_i$ and $b = x_{i+1}$. Our overall symbolic approximation would then be the following quadrature rule.

Simpson's Quadrature Rule.

On each subinterval $[x_i, x_{i+1}]$, evaluate $f(x_i)$, $f(x_{i+1})$, and $f((x_{i+1} + x_i)/2)$. Each function value has weight 1/6, 1/6, and 2/3, respectively, in computing the area under each subinterval. If there are n subintervals, the overall area is

$$A_n = \sum_{i=0}^{n-1} (x_{i+1} - x_i)\left(\frac{1}{6}f(x_i) + \frac{1}{6}f(x_{i+1}) + \frac{2}{3}f\left(\frac{x_i + x_{i+1}}{2}\right)\right).$$

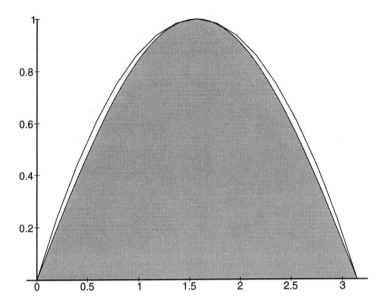

Figure 4.11. Approximating the area under $y = \sin x$ over $[0,\pi]$ by the area under a single quadratic curve.

n	Area (A_n)
1	2.094395
2	2.004559
4	2.000269
8	2.000017
16	2.000001
32	1.999999

Table 4.3. The effect of the number of quadratic curve segments on the accuracy of the definite integral of $\sin x$.

Returning to our attempt to integrate $\sin x$ numerically, Figure 4.11 gives the quadratic approximation to $\sin x$ over a single subinterval $[0,\pi]$, and Figure 4.12 illustrates $\sin x$ decomposed into four quadratic curves. As is apparent, the choice of quadratic curves is particularly appropriate in this case, and the very good results in Table 4.3 bear this out.

The main difficulty with approaches based on Lagrange interpolation is that cusps will usually arise at the joints between curve segments. This is because we have only ensured that curve segments meet at joints but not that their first or

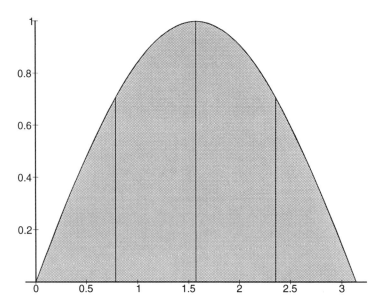

Figure 4.12. Approximating the area under $y = \sin x$ by the area under four quadratic curves.

second derivatives are the same. Figure 4.13 illustrates this behaviour on the function $y = \text{sinc}\, x$. The cusps at segment joints are quite evident. Since for the purposes of integration we are ultimately interested only in the quality of the integral and not the shape of the approximating curve, the effect of cusps may be negligible, and in any case the effect can in practice be reduced by increasing the number of subintervals over which the integral is performed. This results in an increased number of function evaluations; in worst-case situations, this may not be practical, but as we shall see later, Simpson's rule is quite effective. In any case, it is often useful to use piecewise polynomial representations that are inherently smooth across their joints. Most textbooks suggest cubic (or higher-degree) Hermite interpolation. Just to be different, we shall consider the use of cubic Catmull-Rom interpolation. Its use raises some slightly different and interesting issues concerning parameterisation and boundary conditions.

Example 4.8. A graphical estimate to the arc length integrand of $\sin x$ on four curve segments appears in Figure 4.14. An approximation using eight curve segments to the definite integral gives 3.82019780, correct to eight decimal places.

Exercise 4.4. From our work in Chapter 2, derive a quadrature method based on the integration of a four-point Lagrange cubic polynomial.

Exercise 4.5. Implement the procedure **ApproxLagrange(f, [a,b], n, x)** from page 188 in Maple (before looking at Appendix A).

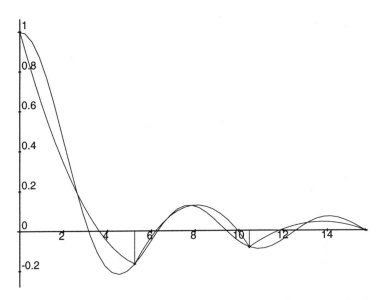

Figure 4.13. Cusps between quadratic curve segments when approximating $y = \text{sinc}\,(x)$ on $[0, 5\pi]$.

4.2.2.3. Smooth Piecewise Cubic Interpolation

We can exploit the machinery we developed in Chapter 2 using the following intuition: if we create a good smooth approximation to a curve, then the integral under our approximate curve should be a good estimate of the real integral. To a large extent, this intuition rings true, but recall that interpolating curves can sometimes be poor approximations to the original function. In general, however, smooth interpolating formulations, when used with care, allow us to make good use of a small set of sample points. Thus this approach is often preferable in such cases when computing sample points themselves is expensive.

In principle, our approach is really no different from that employed to construct the trajectory of a fly from a set of samples. We simply take a sequence of function samples

$$\mathbf{P} \;=\; P_0, P_1, \cdots, P_n \;=\; (a = x_0, f(x_0)), (x_1, f(x_1)), \cdots, (x_n = b, f(x_n)),$$

and pass a set of m curves $C_1, C_2, \cdots, C_m$ through them. There is no requirement that the x_i be equally spaced, but for simplicity we shall make this assumption. If the m curves cover our domain of integration $[a, b]$ and the curves are disjoint, then our approximation is

$$\int_a^b f \;=\; \sum_{i=1}^m \int_{I_i} C_i, \tag{1}$$

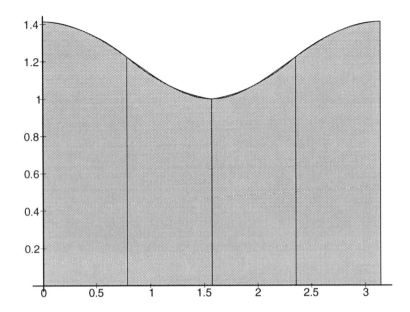

Figure 4.14. Approximating the arc length of $y = \sin x$ by the area under four quadratic curves.

where I_i denotes the interval of integration for each curve segment. Note that I_i is
not necessarily $[x_i, x_{i+1}]$ as in the previous cases, although it can be. There are
three issues that we must address before this approximation scheme can be
implemented:

- What is I_i?
- How can parametric representations for the C_i be handled?
- How can we handle formulations that do not interpolate the first and/or
  last endpoints? In other words, what should our *boundary conditions* be
  on the approximation?

We shall see that the first two issues are highly connected and that the third is
handled differently (but easily) for each curve formulation. To make our
discussion concrete, we shall use as an example our old friend the cubic Catmull-
Rom basis, because it will require us to think carefully about all three issues.

Recalling our discussion in Chapter 2, the cubic Catmull-Rom form on four
points A, B, C, D will result in a cubic polynomial $p(t)$ that varies from B to C
when t varies from 0 to 1. If we subsequently add a point E, then the resulting
polynomial $q(t)$, defined from B, C, D, E, will therefore interpolate C through D.
If we take our $n + 1$ data samples $\mathbf{P}$ defined above, then we would create $n - 2$
curve segments that interpolate all the data except $(a, f(a))$ and $(b, f(b))$, and in
fact would not be defined on the first and final subintervals.

There are many ways around this problem, and most revolve around adding "phantom vertices" that in general are not themselves interpolated but that cause the initial data points to be interpolated:

- Duplicate the first and last data points. Our first curve segment would therefore be derived from input data of the form A, A, B, C, and so the curve would interpolate from A to B as desired. Similarly, the last curve segment would interpolate data of the form A, B, C, C, which again results in endpoint interpolation.

- Choose arbitrary points P_{-1} and P_{n+1} with x values respectively to the left of a and to the right of b.

The first approach has the advantage of requiring no extra function evaluations. While the interval endpoints a and b are interpolated, the tangent to the curve at each endpoint is likely to be incorrect. This could result in excessive overshoot of the function at a or b. The second approach is rather general, and again there are many choices possible (and employed in practice):

- Within the specified constraints, choose an arbitrary point (x,y) near the endpoint. Note that y need not correspond to $f(x)$, so a function evaluation can be avoided.

- Choose a point (x,y) that is on the same line segment as the first two (or respectively the last two) data values. Again, y need not be $f(x)$, and probably will not be.

- Choose a point $(x, f(x))$ at the same spacing to the left of (respectively to the right of) the endpoints.

The first approach can have a behaviour similar to duplicating the endpoints, and consequently exhibits overshoot. The second approach tends to reduce the effect of this problem because the behaviour of the tangent at that point is constrained. The third approach removes the problem altogether, but it requires two extra function evaluations, and the function may not be defined outside $[a,b]$. There are situations in which each of these approaches are usable. When many samples are being taken anyway (i.e., n is large) and when the function is defined outside $[a,b]$, then the third approach is probably more robust and is recommended, since the cost of two extra function evaluations would be small compared to n. This deals with the issue of boundary conditions.

The issue of parameterisation immediately tells us what each subinterval I_i is: in our case, for each curve segment i it is $[0,1]$. However, we have a slight problem here. Since each curve segment is parameterised over $[0,1]$, this amounts to assuming that the size of each subinterval is one rather than some arbitrary value $\Delta x_i = x_{i+1} - x_i$. Dealing with this is simple: after integrating our function C_i (from Eq. 1) over $[0,1]$, scale the result by Δx_i. That is,

$$\int_{I_i} C_i \;=\; \Delta x_i \int_0^1 C_i(t)\,dt. \qquad (2)$$

This relationship is so fundamental that it is worth verifying both in practice and in theory. To integrate a function $f(x)$ on $[a,b]$, we have been breaking up the interval into one or more pieces, taking some samples of f, fitting polynomials of some degree through those pieces, and finding the area of each one. In our case, *each* of these polynomials is parameterised on $[0,1]$, and Eq. 2 relates the integral of the parameterised curves to the actual integral we need to compute. Now, suppose we approximate the integral of $f(x)$ over $[a,b]$ by a single polynomial $p(x)$ over the same interval. That is,

$$\int_a^b f(x)\,dx \approx \int_a^b p(x)\,dx.$$

The polynomial $p(x)$ can be reparameterised to a polynomial $\hat{p}(t)$ such that $t \in [0,1]$. The relationship between x and t should be familiar from Chapter 1:

$$x(t) = t(b-a) + a, \tag{3}$$

and $\hat{p}(t) = p(x(t))$. For the sake of concreteness, let us assume that $p(x)$ is a cubic polynomial. Recalling Exercise 1.42 at the end of Chapter 1, we can use the function **reparam** in Maple to establish Eq. 2 for cubic polynomials, although in fact the argument is independent of degree. We will use colons instead of semi-colons because the expressions are somewhat messy to reproduce; the reader is encouraged to type these expressions in using a semi-colon.

```
> px := A*x^3 + B*x^2 + C*x + D;

                3        2
        px := A x   + B x   + C x + D

> pt := reparam(px, x, [a,b], t):
> Px := int(px, x=a..b):
> Pt := int(pt, t=0..1):

> simplify(Px/Pt);

                b - a
```

Px is the unparameterised approximation to the integral of $f(x)$, while **Pt** is the parametric version. These values differ by exactly $\Delta x = b - a$, so that in fact,

$$\int_a^b p(x)\,dx = \Delta x \int_0^1 \hat{p}(t)\,dt.$$

In calculus, Eq. 3 is called a *change of variables*, which is really a synonym for *reparameterisation*. The behaviour of reparameterised functions under the integral has been thoroughly studied and allows us to be completely general. Let $x:[0,1] \rightarrow [a,b]$ be as in Eq. 3 so that $x(0) = a$ and $x(1) = b$. Also let $p(x)$ and $\hat{p}(t) = p(x(t))$ be as above (though the result is not restricted to polynomials). A basic theorem from calculus[1] states that

---
1. For example, page 301 of J.E. Marsden, *Elementary Classical Analysis*, W.H. Freeman and Co., San Francisco, 1974.

$$\int_a^b p(x)\,dx \;=\; \int_0^1 p(x(t))\,x'(t)\,dt.$$

The right-hand side of the equation looks messy, but in fact observe that $p(x(t))$ is just our parametric polynomial $\hat{p}(t)$. Differentiating $x(t)$ from Eq. 3 with respect to t gives Δx. Therefore

$$\int_a^b p(x)\,dx \;=\; \Delta x \int_0^1 \hat{p}(t)\,dt.$$

If we now break $[a,b]$ up into subintervals $[x_i, x_{i+1}]$, and pass curve C_i through each such subinterval, then we have exactly Eq. 2. This covers all of our considerations.

In summary, we create two "phantom" data points P_{-1} and P_{n+1} that cause all of the original data points P_0 through P_n to be interpolated, giving rise to n curve segments. The final form of the approximation is thus

$$\int_a^b f \;=\; \sum_{i=-1}^{n-2} \Delta x_{i+1} \int_0^1 C_i(t)\,dt,$$

where $\Delta x_i = x_{i+1} - x_i$. Recalling Chapter 2, each curve segment C_i is defined by multiplying together the Catmull-Rom blending functions b_0, b_1, b_2, b_3 and the relevant data points:

$$C_i(t) \;=\; \sum_{j=0}^{3} y_{i+j}\,b_j(t).$$

If our goal were only to plot the curve segments C_i or to compute the integrals symbolically, we could stop here. But our goal is to compute a definite integral efficiently. Noticing that the C_i are themselves sums of polynomials, and that each C_i is individually integrated, we can perform some simple algebra:

$$\int_0^1 C_i(t)\,dt \;=\; \int_0^1 \sum_{j=0}^{3} y_{i+j}\,b_j(t)\,dt$$

$$=\; \sum_{j=0}^{3} \int_0^1 y_{i+j}\,b_j(t)\,dt$$

$$=\; \sum_{j=0}^{3} y_{i+j} \int_0^1 b_j(t)\,dt.$$

What this means is that we can directly integrate the Catmull-Rom blending functions (or those of another basis) as a precomputation. These values then simply become weights for the sample values y_k; once again parameterisation on $[0,1]$ has given us substantial gains. From Chapter 2, recall that the cubic Catmull-Rom blending functions are:

$$b_0(t) = -\frac{t^3}{2} + t^2 - \frac{t}{2},$$

$$b_1(t) = \frac{3t^3}{2} - \frac{5t^2}{2} + 1,$$

$$b_2(t) = -\frac{3t^3}{2} + 2t^2 + \frac{t}{2},$$

$$b_3(t) = \frac{t^3}{2} - \frac{t^2}{2}.$$

Computing the integral of each b_j is very easy. From an exercise above, if

$$b_j(t) = \sum_{i=0}^{3} a_i t^i,$$

then

$$B_j(t) = \int_0^1 b_j(t)\,dt = \sum_{i=0}^{3} \frac{a_i}{i+1}.$$

Therefore

$$B_0 = \frac{-1}{24}, \quad B_1 = \frac{13}{24}, \quad B_2 = \frac{13}{24}, \quad B_3 = \frac{-1}{24}.$$

Gathering everything up, our quadrature rule looks like this.

Cubic Catmull-Rom Quadrature Rule.

Over each subinterval i simulate the effect of integrating the cubic Catmull-Rom curve passing through the points $f(x_i)$ and $f(x_{i+1})$. Doing so requires the values $f(x_{i-1})$, $f(x_i)$, $f(x_{i+1})$, and $f(x_{i+2})$. Each function value has weight $-1/24$, $13/24$, $13/24$, $-1/24$, respectively, in computing the area under each subinterval. If there are n subintervals, then the overall area is

$$A_n = \sum_{i=0}^{n-1} (x_i - x_{i-1}) \left(\frac{-1}{24} f(x_{i-1}) + \frac{13}{24} f(x_i) + \frac{13}{24} f(x_{i+1}) - \frac{1}{24} f(x_{i+2}) \right).$$

Unlike Simpson's rule, some of these weights are negative, but most negative weights vanish, as we shall see. Let us revisit our continuing examples. Figures 4.15 and 4.16 depict Catmull-Rom interpolation of $\sin x$ over $[0,\pi]$ using four and five data points, respectively, on that interval. Figure 4.17 depicts its performance on our arc-length integrand. Observe that the curve both undershoots and overshoots the function, which is natural in higher-degree interpolation. Table 4.4 gives approximations to the integral of $\sin x$ using different numbers of samples.

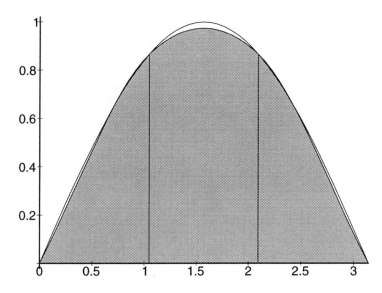

Figure 4.15. Approximating the area under $y = \sin x$ using cubic Catmull-Rom interpolation on three segments over this interval.

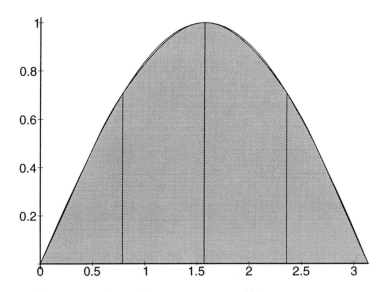

Figure 4.16. Approximating the area under $y = \sin x$ using cubic Catmull-Rom interpolation on four segments over this interval.

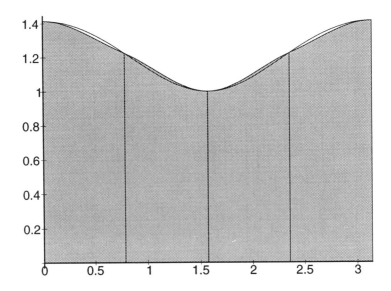

Figure 4.17. Approximating the arc length of $y = \sin x$ by the area under four
Catmull-Rom curve segments.

n	Area (A_n)
3	1.964949
6	1.997730
12	1.999857
24	1.999991
48	1.999999

Table 4.4.   The effect of the number of segments on the accuracy of the definite integral of $\sin x$ using cubic Catmull-Rom interpolation.

Example 4.9.   Using this technique to compute the arc length of $\sin x$ over $[0,\pi]$ gives
3.820197788, correct to eight decimal places, using 32 function evaluations.

Exercise 4.6.   Adapt our discussion above to employ cubic Hermite interpolation instead of
Catmull-Rom. How do the boundary conditions differ?

4.2.3. Formulae for Compound Integration

All of the integration techniques we have discussed reduce to taking a weighted sum of function values scattered on an interval. The approaches were presented in a piecewise setting, but we did not take advantage of the fact that the same function values are often used in more than one subinterval. For efficiency, we could accumulate the weight applied to each sample and write the *compound* integral out as an overall weighted sum. We shall again assume that for each technique, we have at our disposal a sequence of samples over n subintervals

$$(x_0, f(x_0)), (x_1, f(x_1)), \cdots (x_n, f(x_n))$$

with $a = x_0$ and $b = x_n$.

In the case of Riemann sums, the sample value at the midpoint m_i of a subinterval I_i is used exactly once, so the compound formula on n subintervals coincides with our original formula above:

$$A_n = \sum_{i=0}^{n-1} (x_{i+1} - x_i) f(m_i).$$

On the other hand, recalling piecewise linear interpolation or trapezoid rule,

$$A_n = \sum_{i=0}^{n-1} \frac{f(x_{i+1}) + f(x_i)}{2} (x_{i+1} - x_i),$$

the righthand endpoint on subinterval I_i becomes the left-hand endpoint of interval I_{i+1}. If we assume uniformly spaced x, we can rewrite this equation as

$$A_n = \Delta x \sum_{i=0}^{n-1} \frac{f(x_i) + f(x_{i+1})}{2}$$

$$= \Delta x \left(\frac{f(x_0)}{2} + \frac{f(x_1)}{2} + \frac{f(x_1)}{2} + \cdots + \frac{f(x_{n-1})}{2} + \frac{f(x_{n-1})}{2} + \frac{f(x_n)}{2} \right)$$

$$= \Delta x \left(\frac{f(a)}{2} + f(x_1) + f(x_2) + \cdots + f(x_{n-1}) + \frac{f(b)}{2} \right).$$

Compound Trapezoid Rule.

If $x_0, x_1, \cdots, x_n$ are uniformly spaced points on the interval $[a,b]$ with $x_0 = a$ and $x_n = b$, then the compound trapezoid rule on these points is

$$A_n = \Delta x \left(\frac{f(a)}{2} + f(x_1) + f(x_2) + \cdots + f(x_{n-1}) + \frac{f(b)}{2} \right).$$

To derive a compound Simpson's rule, we once again assume that the x_i are uniformly spaced:

$$A_n \doteq \sum_{i=0}^{n-1} \frac{x_{i+1} - x_i}{6} f(x_i) + \frac{x_{i+1} - x_i}{6} f(x_{i+1}) + \frac{2(x_{i+1} - x_i)}{3} f\left(\frac{x_i + x_{i+1}}{2}\right)$$

$$= \frac{\Delta x}{6} \sum_{i=0}^{n-1} f(x_i) + f(x_{i+1}) + 4f\left(\frac{x_i + x_{i+1}}{2}\right)$$

$$= \frac{\Delta x}{6} \Big(f(a) + 4f(m_0) + 2f(x_1) + 4f(m_1) + \cdots + 4f(m_{n-1}) + f(b) \Big)$$

where $m_i = (x_i + x_{i+1})/2$. Note, however, that the sequence of x values

$$x_0, m_0, x_1, m_1, \cdots, x_{n-1}, m_{n-1}, x_n$$

contains $2n + 1$ points with an equal spacing of $\Delta x/2$. We relabel these values as

$$\overline{x}_0, \overline{x}_1, \cdots, \overline{x}_{2n},$$

where $\overline{x}_{2i} = x_i$ for $i = 0, 1, \cdots, n$. and $\overline{x}_{2i+1} = m_i$, for $i = 0, 1, \cdots, n-1$. We also let $\Delta \overline{x} = (x_{i+1} - x_i)/2 = \Delta x/2$. With this we can rewrite the above equation for compound Simpson's rule as

$$A_n = \frac{\Delta \overline{x}}{3} \Big(f(a) + 4f(\overline{x}_1) + 2f(\overline{x}_2) + 4f(\overline{x}_3) + \cdots + 4f(\overline{x}_{2n-1}) + f(b) \Big),$$

since $\Delta x/6 = \Delta \overline{x}/3$. We shall drop the $\overline{x}$ notation in the following rule.

Compound Simpson's Rule.

The result of approximating a function f piecewise on n intervals, using n piecewise continuous quadratic curves and integrating the resulting approximation is

$$A_n = \frac{\Delta x}{3} \Big(f(a) + 4f(x_1) + 2f(x_2) + \cdots + 4f(x_{2n-1}) + f(b) \Big),$$

where $\Delta x = \dfrac{b-a}{2n}$ and $x_i = a + i\Delta x$, $i = 0, 1, \cdots, 2n$.

Exercise 4.7. Show that the compound cubic Catmull-Rom rule on equally spaced data

$$(x_{-1}, f(x_{-1})), (x_0, f(x_0)), (x_1, f(x_1)), \cdots (x_n, f(x_n)), (x_{n+1}, f(x_{n+1}))$$

with $a = x_0$ and $b = x_n$, is

$$\Delta x \left(\frac{-1}{24} f(x_{-1}) + \frac{12}{24} f(x_0) + \frac{25}{24} f(x_1) + \sum_{i=2}^{n-2} f(x_i) + \frac{25}{24} f(x_{n-1}) + \frac{12}{24} f(x_n) + \frac{-1}{24} f(x_{n+1}) \right).$$

This rule looks somewhat like the compound trapezoid rule with a "correction" at the two boundaries.

Example 4.10. The equation in the previous exercise can be written in a more interesting way. Let $y = p(t)$ be the cubic Catmull-Rom curve segment given by points A, B, C, D. Use Maple to verify that

$$p'(0) = \frac{dp}{dt}(0) = \frac{C - A}{2},$$

$$p'(1) = \frac{dp}{dt}(1) = \frac{D - B}{2}.$$

To get the derivative of p as a function of x, we recall from Eq. 3 that for each curve segment i,

$$x(t) = t \, \Delta x_i + x_i,$$

which is just a linear transformation of t. Thus

$$t = \frac{x - x_i}{\Delta x_i},$$

and since Δx_i is a constant, Δx, we see that

$$\frac{dt}{dx} = \frac{1}{\Delta x}.$$

By a simple application of the chain rule,

$$\frac{d}{dx} p(t) = \frac{dp}{dt} \cdot \frac{dt}{dx} = \frac{p'(t)}{\Delta x}.$$

The derivatives with respect to x at each endpoint are thus

$$d_B = \frac{d}{dx} p(0) = \frac{C - A}{2\Delta x},$$

$$d_C = \frac{d}{dx} p(1) = \frac{D - B}{2\Delta x}.$$

These tangents in fact have a special form called the *central difference* approximation for the derivative of f at the knots corresponding to B and C. Given all this, we can rearrange the compound rule for cubic Catmull-Rom quadrature of f on $[a,b]$ as follows:

$$\Delta x \left(\frac{12}{24} f(x_0) + \sum_{i=1}^{n-1} f(x_i) + \frac{12}{24} f(x_n) \right) + \frac{\Delta x^2}{12} (d_a - d_b).$$

This states that compound integration based on cubic Catmull-Rom interpolation has a behaviour like the compound trapezoid rule except for a (potentially significant) "correction" based on an estimate of the tangent to the curve at a and b. Notice that the importance of this correction rapidly diminishes as the number of subintervals increases, or in other words as Δx decreases (why?).

4.2.4. Adaptive Numerical Integration

If we do not already know the value of the integral, how do we know when our approximation is close enough? Since we cannot dwell at length on error analysis at this point, we cannot offer formulae that predict the quality of our approximations. Such formulae do exist, however, and it should come as no surprise that the error is related to the subinterval sizes. For our purposes, we can offer a simple iterative schema that allows us to approximate definite integrals to as much accuracy as our machine will allow, if desired. Suppose the user has specified a tolerance τ to which the approximation must be faithful. There are many ways of measuring error, as will be seen shortly (and in a later chapter), but let us make use of τ as in the following Maple code. Here, we are relying on the use of function **IntCatRom(f,r,N)** which approximates the integral of **f** on **r**=$[a,b]$ over **N** subintervals.

```
#
# Compute integral using CatRom of f over r to tolerance τ
# with n>2 initial samples.  Stop after m iterations if
# no convergence.
#
IntegrateCatRom := proc(f,r,n,m,τ)
    local i1,i2,N,i;
    N   := n;
    i1 := IntCatRom(f,r,N);
    for i from 1 to m do
        N   := N*2;
        i2 := IntCatRom(f,r,N);
        if (error(i1-i2) < τ) then
            RETURN(i2);
        fi;
        i1 := i2
    od;
    ERROR('Integrate: did not converge',i2);
end:
```

This simple-minded algorithm iteratively computes the definite integral until the difference between successive computations is within the threshold τ. It is not efficient because it doubles the number of sample points on each iteration. Another approach would be to refine each subinterval adaptively depending on its "share" of the tolerance that it must achieve.

Exercise 4.8. The points used in iteration i be reused in iteration $i+1$. Recalling Exercise 4.7, rewrite the compound rule to allow this. Then rewrite the above algorithm to reuse values.

4.3. Comparison of Results

We have already said that we do not yet have the mathematical fundamentals to analyse the accuracy of the approaches we have discussed. However, we can compare the results of actual computations to get some idea of their performance. For well-behaved functions, we have already seen that all of the techniques can compute their integrals fairly accurately. How then can the performance of different quadrature routines be assessed? At least two simple measures come to mind:

- How many function evaluations are required to achieve a given tolerance?
- How well can the routine handle less well-behaved functions? This in turn reduces to at least two further questions. First, does the routine give wildly inaccurate results for some functions? Second, do the number of function evaluations go up dramatically?

We consider two simple (and related) quantitative results. Table 4.5 summarises the number of function evaluations required for each approach to meet a specified relative error. The error is measured as follows. If i is a precise value of the integral and a is our approximation, then the *relative error* of a with respect to i is

$$\frac{|i-a|}{i}.$$

As long as i is not very close to zero, relative error is usually a more appropriate measure than *absolute error*, which we define as

$$|i-a|,$$

which of course is simply the numerator of the above expression. Relative error expresses the intuition that if the integral i is a large value, then we are (probably) willing to tolerate greater absolute error. If i is small, on the other hand, then we are likely to be more exacting. Relative error expresses proportional error: if for example $i = 100$ and $a = 99.99$, then the relative error is $0.01/100 = 0.0001$, or 0.01% error.

Table 4.6 illustrates the "bang for the buck" each technique gives. For 100 function evaluations, the relative error achieved is listed. These two tables are at best a very coarse indication of the performance of these techniques, but some trends are evident.

None of the routines have any difficulty with computing the arc length of $\sin x$. As we saw above, this is a very smooth function. On the other hand, the computation of the integral of $\sin x$ and of Runge's function was problematic for the two low-order techniques. The only function that begins to test the mettle of the higher-order techniques is the sinc function. From the data, there is no clear winner between cubic Catmull-Rom interpolation and Simpson's rule, although

Integral	Riemann	Trapezoid	Simpson	Catmull-Rom
sine	2048	2048	53	60
arc length	7	10	10	10
Runge	575	750	53	50
sinc	2100	2000	107	128

$$\text{sine: } \int_0^\pi \sin x \, dx$$

$$\text{arc length: } \int_0^\pi \sqrt{1 + \cos^2 x} \, dx$$

$$\text{Runge: } \int_{-1}^1 \frac{1}{1 + 25t^2} \, dt$$

$$\text{sinc: } \int_0^{5\pi} \frac{\sin \pi x}{\pi x} \, dx$$

Table 4.5. Various definite integrals and the number of function evaluations required to achieve a relative error of 0.1×10^{-6}.

Integral	Riemann	Trapezoid	Simpson	Catmull-Rom
sin	0.4112×10^{-4}	0.8742×10^{-4}	0.6000×10^{-8}	0.1550×10^{-7}
arc	0.7853×10^{-9}	0.1047×10^{-8}	0.2356×10^{-8}	0.1047×10^{-8}
Runge	0.3520×10^{-6}	0.6980×10^{-6}	0.2986×10^{-8}	0.6635×10^{-9}
sinc	0.4009×10^{-4}	0.8518×10^{-4}	0.1555×10^{-6}	0.3947×10^{-6}

Table 4.6. The relative error made using 100 function evaluations.

there seem to be two observable trends. The first is that cubic interpolation may deliver slightly better performance when only a small number of function samples can be used. The second is that Simpson's rule overcomes this when the computation is taken further to meet a lower tolerance. An explanation may be that the extra smoothness (i.e., lack of cusps) afforded by cubic Catmull-Rom interpolation pays dividends when only a small number of samples is used, but that as the number of samples increases, the arcs in the function being approximated are increasingly straight. Thus a cubic approach may wiggle more than the quadratic approach, thus incurring greater error. This explanation, while plausible, has not been carefully analysed.

As for efficiency, the Catmull-Rom approach is slightly faster, because it saves on a division. This efficiency is solely due to the fact that because all curve segments are parameterised on [0,1], we can precompute weights using the parametric approach.

Exercise 4.9. Can Simpson's rule (and piecewise Lagrange interpolation of higher orders) be parameterised in the same way as our smooth piecewise cubic formulations?

4.4. Monte Carlo Methods

A great many pubs around the world have a games room in which one can play a game of darts among friends and savour an occasional refreshment. It is hard to imagine that such a cheerful setting can provide a compelling demonstration of a completely different kind of quadrature.

Most dartboards worth their salt are circular. Suppose that among all the good cheer, someone asks what the area of the dartboard is. This sort of question comes up often enough in a pub, but somehow we have all forgotten our geometry. Nonetheless, we would like to give an approximation of this area. Here is what we could do. We cut out a square of cardboard or cork so that that it is big enough to completely surround the dartboard. Despite our geometric memory lapse, we manage to compute the area of this square, and we write it down just to be sure. Let this area be A. We paste the cardboard in behind the dartboard and we begin to throw darts.

Now, in the interests of mathematical truth, we should not throw our darts like experts. Indeed, we must throw them quite haphazardly, with the only constraint being that we should try to hit the surrounding cardboard for fear of damaging the wood finish on the wall. After about a thousand dart throws, our dartboard together with the surrounding square might look something like Figure 4.18. After a while we begin to gather an admiring crowd, but we begin to get bored with the proceedings. At that point, we count the number of dart impressions on that dartboard and divide by the total number of darts thrown. We multiply this ratio by A, and we confidently report a fairly accurate value for the area of the circle. The C program in Figure 4.19 simulates in a much less colourful way what we have just done.

After 1 000 throws, the program reports an area for the unit circle that is correct to 1 decimal place, namely 3.1. After 100 000 throws, it is correct to two places: 3.14. After 10 000 000 throws, it reports 3.141. Notice that we are getting about one extra decimal place of accuracy by taking 100 times as many samples. We shall return to this point later.

The experiment we have performed is one example of a *Monte Carlo* method for computing an integral. Notice that the actual code to do the integration in the above C code is on the order of seven lines. Ease of implementation is one of the

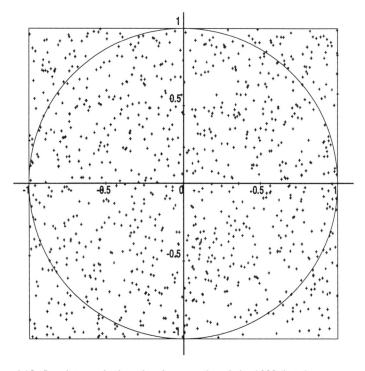

Figure 4.18. Imprints on dartboard and surround made by 1000 dart throws.

major attractions of Monte Carlo methods. But one would hardly think this would be an efficient way to compute integrals. For integrals over domains of low-dimension, this is certainly the case, for there are many good ''deterministic'' quadrature schemes that can easily handle such problems. However, as the dimensionality of the domain increases, most deterministic schemes begin to fail, until at some point all that remains are techniques based on point or area sampling. It may seem astonishingly expensive to take 10 000 000 samples of a function, but consider this analogy. On a 1-D domain, it is rarely sufficient to take five samples of a function to compute its integral, whether the samples are randomly or regularly placed. Over a 2-D rectangular domain, a 5×5 grid of values, namely 25 samples, just might be satisfactory; in 3-D, a 5^3 grid might suffice. Continuing along this road, we see that to maintain even a dubious amount of accuracy over a 10-dimensional domain, we would probably require a grid of size at least 5^{10} or almost 10 000 000 samples. In high-dimensional problems, a great number of samples do not go very far, especially if they are regularly spaced. Amazingly, the statistical analysis of Monte Carlo methods shows that their accuracy is actually *independent* of dimension and depends primarily on the number of samples taken and their distribution.

```
/*      Compute area of unit circle using dart throwing.  */
long random();
#define Rand01 ( (double)random()/(maxrand) )  /* a random number in [0,1] */

main()
{
        double      sum, u,v;
        long int    nmax = 10000000;       /* a big set of darts */
        long int    n;
        int         i;
        double      f(), pow();
        double      maxrand;               /* largest random number */

        maxrand  = pow(2.0,31.0)-1;        /* maxrand = 2^31 − 1 */
/*      For each order of magnitude of n, throw n darts.    */
        for (n=10; n<=nmax; n*=10) {
            sum = 0.0;
            for (i=0; i<n; i++) {
                u   = 2.0*(Rand01 - 0.5);
                v   = 2.0*(Rand01 - 0.5);
                sum  = sum + f(u,v);
            }
/*          Now we have the proportion of the area covered.     */
/*          Multiply by the area of the square to get our estimate. */
            sum = 4.0*sum/(double)n;
            printf("estimated area of circle on %d samples: %f.",n, sum);
        }
}

double f(double x, double y)        /* is (x,y) inside the circle? 1=yes, 0=no */
{
        if(x*x + y*y <= 1.0)
            return(1.0);
        else
            return(0.0);
}
```

Figure 4.19. A program to compute the area of a unit circle using 10, 100, ..., 10,000,000 darts.

Let us take a step back and describe Monte Carlo methods more carefully. Space does not permit us go much further than to give a hint of the statistics underlying the method. Nevertheless, we can still acquire a good, nonrigorous idea of the mechanics of Monte Carlo methods.

Suppose our goal is to estimate the integral of $f: \mathbf{R} \to \mathbf{R}$ on $[a,b]$:

$$I = \int_a^b f(x)\,dx.$$

We begin by choosing an arbitrary (and probably large) number of random values $U = u_1, u_2, \cdots, u_n$, all in $[0,1]$. Then just as we would compute the average of our grades by summing them and dividing by the number of grades, we would naturally expect the average or *mean* of the $u \in U$ to be

$$\bar{u} \;=\; \frac{1}{n}\sum_{i=1}^{n} u_i \;\approx\; \frac{1}{2}.$$

We are assuming that the *distribution* of random values is *uniform*. It is easier to understand what this means in a discrete setting. For example, if a die (the singular of *dice*) is fair, then the likelihood of rolling a 2 is the same as that of rolling a 1 or a 6—namely one chance in six, or 1/6. That is, there is equal probability of rolling any number from 1 through 6. When dealing with continuous domains such as real numbers on the interval $[0,1]$, the notion of "equally probable" must be defined more carefully, but hopefully the reader sees the analogy.

An optional remark on continuous probability distributions. Imagine having a very thin dart that we can throw on a continuous domain such as $[0,1]$. The probability of the dart covering any specific spot $u \in [0,1]$ is related to the proportion of the width of the dart covering $[0,1]$. As we decrease the width of the dart, we are decreasingly likely to hit a specific point u. In the limit, the chance that we can hit a specific point u with an infinitesimally thin dart is zero. For this reason, the characterisation of a distribution on a continuous domain is usually given by the probability with which subsets of that domain arise. For example, for a uniform distribution on $[0,1]$, the probability that $u = 0.3$ is zero, but the probability that u lies between d and e, $d<e$, in $[0,1]$ is always $e-d$, regardless of the choice of d and e.

We have often seen that we can relabel a value from $u_i \in [0,1]$ to a value x_i in an arbitrary interval $[a,b]$ by the transformation

$$x_i \;=\; a + (b-a)u_i.$$

Since the x_i are shifts and scales of uniformly distributed u_i, the x_i are also uniformly distributed random numbers on $[a,b]$, and their mean is

$$\bar{x} \;=\; a + (b-a)\bar{u} \;=\; a + \frac{(b-a)}{2} \;=\; \frac{(b+a)}{2},$$

as we would expect. Now, consider evaluating f at each x_i. Then we can easily compute the mean of f with respect to these values:

$$\bar{f} \;=\; \frac{1}{n}\sum_{i=1}^{n} f(x_i).$$

As a one-dimensional function, we can think of $\bar{f}$ as the average height of f with respect to the set of uniformly distributed values given by the x_i. Let us denote the real mean of f by $E(f)$, for the *expected value* of f. This does not mean we know what this value is or that it is even computable. As n grows larger, we would expect $\overline{f}$ to be an increasingly good approximation to $E(f)$. If we actually had $E(f)$ available, then

$$I = (b-a)E(f)$$

would give us the integral of f. Because we have an approximation $\overline{f}$ to $E(f)$, we therefore have an approximation for our integral I:

$$I \approx (b-a)\overline{f} = \frac{(b-a)}{n} \sum_{i=1}^{n} f(x_i).$$

This equation is the cornerstone of most Monte Carlo methods. The two Maple procedures in Figure 4.20 depict implementations of this equation.

```
#
#     Monte Carlo integration of 1-D function f
#     on interval I = [a,b] using n samples.
#     Sample invokation: montecarlo(sin,[0,Pi],1000);
#
montecarlo := proc(f, I, n)
     local a,b,          # endpoints of interval
           i,j,          # indices
           randno;       # array of random numbers

     a := I[1]; b := I[2];
     randno := seq( evalf(a + (b-a)*rand()*10^(-12)), i=1..n);
     fval   := map(f, [randno]);
     evalf( (b-a)/n * sum(fval[j],j=1..n));
end:

mc := proc(f, I, n)         # invocation: mc(sin(x),[0,Pi],1000);
     local x,               # indeterminate of f
           a,b,             # endpoints of interval
           u,i;

     x := indets(f,name)[1];
     us := 0;
     s := 0;
     a := I[1]; b := I[2];
     for i from 1 to n do
          u := evalf(a + (b-a)*rand()*10^(-12));
          s := s + evalf(subs(x=u,f));
     od;
     evalf( (b-a)/n * s);
end:
```

Figure 4.20. A Maple program to do 1-D integration. The "clever" Maple routine **montecarlo** runs about two orders of magnitude more slowly than the "dumb" routine **mc**. This is because as **n** grows, the size of the sequences grows. Warning: Maple's **rand()** function at the time of writing is a poor pseudo-random number generator.

It is worthwhile to provide some intuition as to why we seemed to get one extra decimal place of accuracy for 100 times as many samples. In fact, it comes directly from a statistical interpretation for $\bar{f}$. Notice that $\bar{f}$ is actually the sum of n random values multiplied by some constants. Now, the sum of a large number of uniformly distributed random numbers is also random, but is actually not itself uniformly distributed. Again, a discrete example will help. If we roll two dice repeatedly and consider their sum on each roll, then we will notice that 2 and 12 come up least often, followed by 3 and 11, 4 and 10, and so on. If we were to count up the number of frequency of each possible outcome, and plot the result, we would get a triangular distribution as depicted in Figure 4.21.

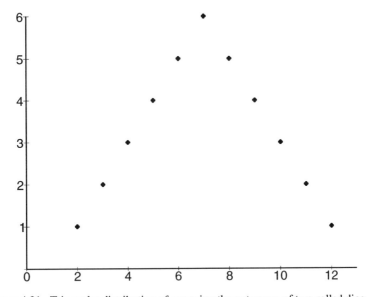

Figure 4.21. Triangular distribution of summing the outcomes of two rolled dice.

If instead we were to roll three dice repeatedly and compute the distribution, we would get a piecewise quadratic bump as in Figure 4.22. As the number of dice we roll increases, we would get a bell-shaped curve, which is in fact our friend the gaussian distribution that we discovered in the last chapter. Indeed, we should not be at all surprised by the shape of these curves: they are in fact exactly the curves we got from iterated moving-average filters in the last chapter. Returning to $\bar{f}$, with each instance $X = x_1, \cdots, x_n$ of uniformly distributed samples, we would get a slightly different $\bar{f}$. If we were to plot the distribution of $\bar{f}$ over many "dice rolls" of X, we would get a gaussian curve. The $\bar{f}$ would therefore be *gaussian distributed*, or *gaussian random variables*. Furthermore, as we increase n, it can be shown that the *variance* σ_n^2 of $\bar{f}$ about the real mean value of f, $E(f)$, decreases in the ratio $1/n$. In other words,

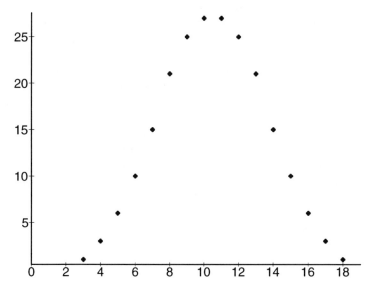

Figure 4.22. Distribution resulting from summing the outcomes of three rolled dice.

$$\sigma_n^2 \;=\; O\!\left(\frac{1}{n}\right).$$

A statistical measure of the error is given by the *standard deviation*, σ_n of the distribution, which is simply the square root of the variance. Therefore,

$$\sigma_n \;=\; O\!\left(\frac{1}{\sqrt{n}}\right).$$

Thus we have inverse square-root convergence. While we have not provided a rigorous statistical argument, we at least have a glimpse of how the argument would work. Further, this provides an explanation of the observed convergence behaviour above, a behaviour that can be summarised as: *to halve the error, we must quadruple the number of samples.*

The 1-D Monte Carlo method we described generalises very easily to an *n*-dimensional domain. Similarly the random-sampling method for computing the areas of objects, as we did for a circular dartboard above, extends both to high dimensions and to arbitrary shapes and volumes, as long as it is possible to determine what points are inside or outside. However, the observed inverse square-root convergence is a difficult problem to overcome and is an area of considerable research activity. One hint that we might be able to do better than $1/\sqrt{n}$ convergence can be found by looking at Figure 4.18 once again. Notice that while the samples chosen appear to cover the region, there are ''clumps'' of samples locally clustering together. Intuitively we would think that one could still have a random sampling but one that doesn't exhibit as much clumpiness. One way to ensure this is to insist that when a sample x_i is chosen, then we are not allowed to choose another sample x_j that comes within distance ε_i of x_i. In two dimensions, this means that each x_i can be thought of as representing a small disk centred at x_i. This sampling rule is known as *Poisson-disk* sampling, and it has many applications in Monte Carlo algorithms and in random sampling for audio and video signals. Notice that because Poisson-disk sampling guarantees a proximity rule among samples, then a point sample can be used to represent an area. In our dartboard example above, we claimed we did not know the area of a circle, so we would not be any better off knowing that each sample was itself a small circle, but in general, we are much better off with such problems using Poisson-disk sampling. One difficulty with the approach, however, is that it can be slow to compute a large set of such samples.

Exercise 4.10. Write a program to compute n Poisson-disk samples $x_1, \cdots, x_n$ over a fixed region such that each sample is a disk of radius ε. Let n grow large and report the running time of the program. Now for several fixed values of n, let ε vary and report the running times.

Exercise 4.11. Compare the number of samples needed to compute the area of an ellipse given by

$$\frac{x^2}{a^2} + \frac{y^2}{b^2} = 1$$

correct to three decimal places. Use Maple (or your own know-how) to compute the areas of various ellipses for reference purposes. Does the result depend on ε?

Exercise 4.12. Use Poisson-disk and uniform sampling to compute the integrals of $y = x^2$, $y = 1/x$, and $y = \ln x$ on the interval $[0,1]$. Once again report the results correct to three decimal places and report the number of samples needed for several different values of ε.

Exercise 4.13. (Difficult but rewarding.) Poisson-disk sampling tries to choose uniformly distributed samples that display better separation and packing behaviour. Suppose we have a fixed region, say a unit square, and a given ε, and we wish to pack the largest possible number of samples into the unit square such that no sample is within ε of any other. What would this sampling pattern be? Hint: is it random? Discuss.

Poisson-disk sampling is often superior to uniform sampling because, as indicated in the previous exercise, it favours sets of uniformly distributed samples that have superior spatial properties. This opens up a deep and fascinating question: do there exist sampling patterns that for a large set of functions exhibit better than inverse square-root convergence? The answer seems to be yes: There appear to be sampling arrangements that are "quasi-random" and that exhibit $1/n$ convergence rather than $1/\sqrt{n}$. This is an significant improvement, but the mathematics of this work is very challenging. What is not yet clear is whether it is practical to generate such sampling patterns. We had a hint of this problem in an exercise earlier when we considered the cost of computing Poisson-disk sampling arrangements. A similarly difficult problem awaits us here.

Exercise 4.14. Use Maple to plot $1/n$ and $1/\sqrt{n}$ to see why the possibility of such sampling arrangements is so exciting.

4.5. Summary

We have considered several approaches to computational integration, including symbolic integration and four numerical quadrature schemes. In general, our approach consisted of scattering a set of samples throughout an interval, and then weighting each sample in some way to derive an approximation to the definite integral. None of the quadrature techniques we discussed are sufficiently robust and efficient to be practical, but variants of them are the basis for many real implementations. Unlike our approaches, some better quality quadrature routines try to be more economical in their use of function evaluations by spacing samples adaptively or nonuniformly; other routines use regularly spaced samples and basis functions that have provably better approximations. This raises the more general issue of allowing both the weighting of the sample points and their placement to vary. Using suitable optimisation criteria, one approach that chooses both the sample spacing *and* the weighting applied at each sample is called *gaussian quadrature*.

So far, we have omitted from our discussion a consideration of both error and convergence. As a first glimpse of the subject, we have looked at the empirical behaviour of our quadrature routines. After discussing series representations, we will be in a better position to revisit this topic.

Another omission is that we have ignored the numerical computation of improper integrals. As it stands, we can compute, with moderate efficiency, the definite integrals of a wide variety of functions. However, if we let either endpoint of the interval tend to infinity, we are unable to solve the problem with any of the above techniques, including Monte Carlo methods. There are two approaches to coping with improper integrals.

One approach is to break the domain of integration up into pieces over definite intervals, and to estimate using other techniques the behaviour of the function in the subinterval that tends to infinity. Often, we try to reason that we can choose this subinterval so that it has negligible effect on the accuracy of the solution. This is sometimes called improper integration by *truncation*.

Another approach is to perform a change of variables. We have often used changes of variable in several ways, but so far they have been linear transformations of the form $x = a + (b - a)u$. If in this case u tends to $\pm\infty$, then so will x. Thus we would have to find a transformation that converts an infinite domain for u into a finite domain for x. See the exercises below.

When truncation on its own fails, a final approach is to modify our quadrature rules so that the spacing between samples grows as we tend to infinity. This technique together with truncation can yield good approximations to an improper integral even when the function does not quickly converge to zero as it approaches $\pm \infty$.

Exercise 4.15. Comment on whether or not the integral

$$y = \int_0^{10} \frac{1}{x}\, dx$$

can be handled using one of our approaches. Make a plot of the integrand as part of your argument. It is also interesting to plot the integral equation

$$y(t) = \int_t^{10} \frac{1}{x}\, dx$$

as t approaches 0.

Exercise 4.16. Comment on the feasibility of computing the following definite integral:

$$y = \int_0^1 \sin \frac{1}{x}\, dx.$$

Again, plot the integrand and the integral for various subintervals of [0,1].

Exercise 4.17. Suppose we wish to compute an integral of the form

$$I = \int_0^\infty f(u)\, du.$$

We wish to make a substitution for u so that the improper integral can be converted to a definite integral on [0,1]. Two good candidates for this are

$$u = -\ln x, \quad \text{and}$$

$$u = \frac{x}{1-x}.$$

Use Maple to explore the effect of substituting an expression in x for u in the above integral. How are the endpoints of the original interval mapped to [0,1]? Hint: you have to be careful about the direction of integration. Use one of our quadrature rules to compute the value of

$$I = \int_0^\infty \exp(-u^2)\, du$$

using both changes for the variable u.

Appendix A: Maple Code to Model Quadrature Rules

The following code is a fairly substantial application of Maple that was used to produce many of the figures in this chapter. After a number of utility routines, the procedures **ApproxStep**, **ApproxSimpson**, and **ApproxCatRom** generate plots corresponding to Riemann sums, Simpson's Rule, and a quadrature rule based on cubic Catmull-Rom interpolation.

```
with (plots):

#
#    Combine two lists [a,b,c, ...], [A,B,C, ...] into
#    ordered pairs [ [a,A], [b,B], [c,C], ... ].
#
pairs := proc(x,y)
    local i;
    [x[i],y[i]] $ i=1..nops(x):
end:


#
#    Compute n+1 equally-spaced samples of f on interval r.
#
MakeSamples := proc(f,r,n)
    local dx, dr, x, y, i;
    dr := r[2] - r[1];
    dx := dr/n;
    x := [r[1]+i*dx $ i=0..n];
    y   := map(f,x);
    [x,y];
end:


#
#    As above, except take one extra sample to left and right
#    of interval.
#
MakePhantomSamples := proc(f,r,n)
    local dx, dr, x, y, i;
    dr := r[2] - r[1];
    dx := dr/n;
    x := [r[1]+i*dx $ i=-1..n+1];
    y   := map(f,x);
    [x,y];
end:
```

```
#
#    Plot n points on curve over interval r.
#
RenderCurve := proc(curve, r, n)
    local i, dt, X;
    dt := (r[2]-r[1])/n;
    X := indets(curve);
    [eval(r[1]+i*dt), eval(subs(X[1]=eval(r[1]+i*dt), curve))]
            $ i=0..n;
end:


#
#    Construct graphical representation of Riemann Sums.
#    Arguments: f is function to be plotted, r=[a,b] is the
#               interval, and n is the number of panels.
#
ApproxStep := proc(f,r,n)
    local dr, dx, i, x, y, m, b,xx, plotf, plotb, t;
    dr := r[2] - r[1];
    dx := dr/n;
    x  := [evalf(r[1]+i*dx) $ i=0..n];
    xx := [(x[i]+x[i+1])/2 $ i=1..n];
    y  := map(f,xx);
    b := x[1],0;
    for i from 1 to n do
        b := b,x[i],y[i], x[i+1],y[i], x[i+1],0;
    od;
    plotf := plot(f,(r[1])..(r[2]), colour=GREEN);
    plotb := plot({[b]}, colour=CYAN,style=line);
    display({plotb, plotf});
end:


#
#    Plot a quadratic curve through each subinterval
#    (Simpson's Rule).
#
ApproxSimpson := proc(f,r,n)
    local i, samples, m, s, a, b, c, x,y, X, Y, t,
            plotf, plota, plotc;
    samples := MakeSamples(f,r,n);
    m := nops(samples[2]);
    b := pairs(samples[1],samples[2]);
    x := samples[1];
    y := samples[2];
    c := NULL;
    for i from 1 to m-1 do
        X := [ x[i], (x[i]+x[i+1])/2, x[i+1] ];
```

```
    Y := [ y[i], f(X[2]), y[i+1] ];
        curve := interp( X,Y, t);
        c := c, RenderCurve(curve, [x[i],x[i+1]], 16),
            [x[i+1],0]; # adds a vertical line
    od;
    plotf := plot({f},(r[1])..(r[2]), colour=GREEN);
    plotb := plot({b},(r[1]-0.1)..(r[2]+0.1), colour=CYAN);
    plotc := plot({[c]}, colour=YELLOW);
    display({plotc, plotf, plotb});
end:


#
#    Cubic Catmull-Rom basis
#
CR   :=   [-1/2*t^3 +      t^2 - t/2,
        3/2*t^3 - 5/2*t^2 + 1,
       -3/2*t^3 +    2*t^2 + 1/2*t,
        1/2*t^3 - 1/2*t^2]:


#
#    Symbolic dot product of x and y.
#
dot := proc(x,y)
    local i;
    sum(x[i]*y[i], i=1..nops(x)):
end:



#
#    Plot a Catmull-Rom Interpolation to curve over
#    interval r using n points.
#
#    This is trickier than Lagrange interpolation because
#    a reparameterisation of each curve segment is needed.
#    The beginning and end of each curve segment includes
#    points on the x-axis so that vertical lines are drawn
#    to denote segment boundaries.
#
ApproxCatRom := proc(f,r,n)
    local i, j, samples, m, M, s, a, b, c, x,y, X,Y, t,
            plotf, plota, plotb;

    samples := MakePhantomSamples(f,r,n);
    M := nops(samples[2]);
    m := M-3;
    s := (r[2]-r[1])/m;

    x := samples[1];
```

```
   y := samples[2];

     a := NULL;
     for i from 1 to m do
         Y := [ y[i], y[i+1], y[i+2], y[i+3] ];
         curve := dot(CR,Y);
         c := RenderCurve(curve, [0,1], 16);
         c := [c[1][1]+(i-1), 0],
              [c[j][1]+(i-1), c[j][2]]$ j=1..nops([c]),
              [c[nops([c])][1]+(i-1), 0];
         a := a, c;
     od;
     i := 'i';
     a := [ [a[i][1]*s-abs(r[1]),a[i][2]] $ i=1..nops([a])];
     plotf := plot({f},(r[1])..(r[2]), colour=GREEN);
     plota := plot({a}, colour=YELLOW);
     display({plota, plotf});
end:


#
#    Sample sessions.
#
arc := x -> sqrt(1+cos(x)^2):# arc length of sin(x)
runge := x -> 1/(1+25*x^2):

ApproxStep(sin, [0,Pi], 2);          # Figure 4.4
ApproxStep(sin, [0,Pi], 8);          # Figure 4.5
ApproxSimpson(sin, [0,Pi], 1);       # Figure 4.11
ApproxSimpson(arc, [0,Pi], 4);       # Figure 4.14
ApproxCatRom(sin, [0,Pi], 3);        # Figure 4.15
ApproxCatRom(runge, [-1,1], 23);     # Figure 4.1 (roughly)
```

Chapter 5
Series Approximations

The representation and approximation of a complex mathematical object by means of a series of simpler components is extremely common and useful. If the components of a series are sufficiently simple, then the representation may be more efficient with which to work, and may offer a deeper understanding of the properties of the object. In this chapter, we shall examine several classes of series representations and consider some of their applications. We first discuss the nature of the computational representation of numbers, and their interpretation as series. We then go on to discuss the symbolic representation of functions using Taylor series and Fourier series. Taylor series provide a useful way of approximating well known functions, but their main utility is in the analysis of error. Specifically, we shall use Taylor series to analyse the error of three quadrature algorithms introduced in the last chapter. Our discussion of Fourier series, among other things, will help us to characterise the elusive notion of the ''frequency'' of a function alluded to in Chapter 3.

5.1. Representations for the Real Numbers

To begin to understand how we might try to represent some numbers $x \in \mathbf{R}$, we first consider a motivating example. In Chapter 2, we discussed the use of Lagrange interpolation for approximating curve trajectories. Figure 5.1 contains two routines that do just that. Both routines compute a sequence of $n + 1$ equally-spaced knots and then evaluate at these knots the function to be approximated. A nth-degree Lagrange polynomial that interpolates these points is then computed. The two routines differ in only one respect: **ApproxLagrange** keeps the computation symbolic, while **ApproxLagrangef** employs floating-point numerical computation.

```
#
# Approximate f(x) over interval r using n equally spaced points.
# Returns Lagrange interpolating polynomial as a function of t.
#
# ApproxLagrange uses symbolic form.
# ApproxLagrangef uses floating-point form.
#
ApproxLagrange := proc(f,r,n,t)
     local dx, x, y;
     dx := (r[2] - r[1])/n;
     x  := [(r[1]+i*dx) $ i=0..n];
     y  := map(f,x);

     sort(collect(interp(x, y, t),t));
end:


ApproxLagrangef := proc(f,r,n,t)
     local dx, x, y;
     dx := evalf( ((r[2] - r[1]))/n );
     x  := [evalf(r[1]+i*dx) $ i=0..n];
     y  := map(f,x);

     sort(evalf(interp(x, y, t)));
end:
```

> Figure 5.1.   Two approaches to Lagrange interpolation. The first is symbolic and
> the second is numeric.

It is instructive to use these two routines to approximate a well-known function. The function $\sin x$ for $x \in [-\pi, +\pi]$ is a good candidate. In Figure 5.2, the polynomial **p1** is the result of symbolic computation, while **p2** is the result of numeric computation. These results are clearly different. Which is the better approximation? We can get a hint of this by examining the evidence.

Since $\sin 0 = 0$, we expect a proper Lagrange interpolation of samples to pass through the origin. A necessary condition for a polynomial

$$\sum_{i=0}^{n} a_i t^i$$

to pass through the origin is that a_0 must be 0.[1] Furthermore (and this is less obvious), an odd function must have no even powers of t, and analogously for an even function. At the regular visual scale of $[-\pi, +\pi]$, both approximations are very close. However, if we plot **p1** and **p2** for values close to the origin, the

---

1. Recall Exercise 2.4 on page 77 in Chapter 2.

difference between them becomes obvious. See Figure 5.3. As an aside, this figure also demonstrates that for values x close to zero, $\sin x \approx x$, since the plot of $\sin x$ at this scale looks like the line $y = x$. A plot of the difference **p1** - **p2** can be found in Figure 5.4. On all counts, it appears that the symbolic version is a more accurate representation of the Lagrange interpolation on six points.

```
> r := [-Pi,Pi];
              r := [- Pi, Pi]

> p1 := ApproxLagrange(sin, r, 5, t):
> p2 := ApproxLagrangef(sin, r, 5, t):

> evalf(p1);
            5           3
     .00597 t  - .1600 t  + .9977 t

>evalf(p2);
          5        -10  4           3      -9  2                        -9
   .00597 t  - .5*10   t  - .1600 t  + .5*10   t  + .9977 t  + .7243*10


> p := p1-p2:

> evalf(p);
          -10  5        -9  3          -8          -10  4        -9  2
   - .24*10   t  + .5*10   t  - .23*10   t + .5*10   t  - .5*10   t

            -9
   - .72*10
```

Figure 5.2. Difference between the symbolic and numerical approaches.

This example does *not* permit us to conclude that symbolic computation is inherently more accurate than numerical computation. However, a rule of thumb is that it is generally wise to keep objects in symbolic form for as long as is feasible. In a symbol manipulation system such as Maple or Mathematica, symbolic objects can be manipulated exactly; the price paid is that such objects typically require much more storage than their numerical approximation, as we shall soon see. Another conclusion we can draw is that numerical evaluation on computers is prone to error, and if our application is sensitive to such errors, we must be very careful to cope with these problems. Again, we stress that this does not imply a value judgement about the use of numerical computation. It is a tool that we must learn to use effectively. The study of *numerical methods* is the science of developing general, robust, and efficient solutions to scientific problems using finite precision, and the use of analytic techniques to demonstrate the accuracy of computations. We shall now consider the basic issues underlying the representation of real numbers in computers and discuss both their benefits and shortcomings.

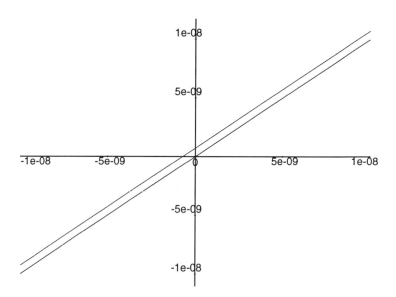

Figure 5.3. The behaviour of **p1** and **p2** near the origin. The one computed using symbolic means passes through the origin.

5.1.1. *The Representation of Integers and Fixed-Point Numbers*

While it is often possible to ignore the technical aspects of a computer representation, especially for logical or string-oriented representations, an awareness of a numeric representation is essential to understanding why things go wrong and when to expect it. A more positive way to say this is that under some conditions, we may be able to guarantee that a numerical method works. Before we can discuss the representation of real numbers, we must first acquaint ourselves with the representation of simpler objects.

A numerical *representation* is a code or language that can be translated into a numerical *value* of interest to us. Just as the same word in spoken or written language may mean different things in different contexts, so too the same code may represent different values depending on its use. A representation should be *expressive*, in that the code together with the translation should allow us to denote most of the values that we are likely to want. We would like to be able to say that a representation is *complete*, in the sense that all possible desired values are representable. For infinite sets, a representation consisting of a finite description is likely to be incomplete. A representation should also be *concise*. Lastly, a representation should be *efficient*, because we wish to be able to encode and decode values quickly. Most often, this will mean that we will implement

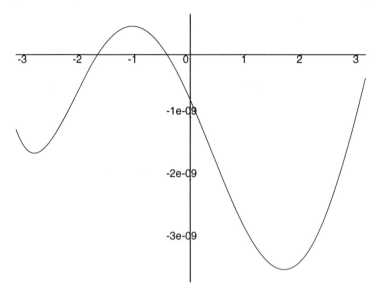

Figure 5.4. A plot of **p1** - **p2**.

operations on values in terms of operations on an encoding of values. This is a fundamental point that is easy to overlook. For example, we are fairly unanimous in our definition of arithmetic addition. There however may be several different implementations of addition in a computer system: one for each numerical representation of interest (not counting forms of addition that can operate on more than one representation).

Representations for use by computers must be built from a basic unit called a *bit*, which represents two possible values: yes/no, black/white, on/off, etc. By cascading bits together, a larger set of values can be represented: putting two bits together allows four values (00, 01, 10, 11), three bits give eight values, and in general k bits give 2^k possible values. The first step in a representation is thus to determine the number n of values we wish to encode. We will thus require about $\log_2 n$ bits to encode these values. The number n is also influenced by pragmatic hardware factors. Since the *byte* or (typically) eight bits is so central to the design of hardware, there is strong motivation to design an encoding scheme that bounds n from above by 2^{8b} for some $b>0$. The number b would then be the number of bytes needed for the encoding. Many hardware designs superimpose the notion of a *word* as well. A word is a unit of operation in a computer's processing unit or a unit of transmission from memory to other parts of the computer. A typical word size is four bytes for transmission and integer arithmetic operations. Computers primarily intended for scientific computations often have larger word sizes. There is no requirement that our representations use full words. It is simply that the cost of working with representations occupying partial words may be as high as those that occupy full words. The notions of a bit, a byte, and a word, which all amount to collections of bits, are fundamental units of data storage in a computer that are difficult to ignore in a representation.

There are various ways to represent numbers in computers, but the fundamental design choice is whether a representation is to be fixed or variable in size. If we choose a representation that always occupies a fixed number of words, then we can design hardware to manipulate these objects very quickly. On the other hand, the actual quantity of distinct numbers that can be represented is relatively small. If we choose instead to allow the representation to vary in size, then a richer set of numbers is expressible, and perhaps more accurately. The disadvantage is that computing with them can take more time and storage.

5.1.1.1. Integers

The details of representing integers is more properly discussed in a course on computer architecture, but we shall briefly motivate three issues:

- that if the integers fall within a representable range, then they can be represented *exactly*.
- that if operations are performed using integers, in some cases the result lies outside the representable range.
- that the representation for integers is an encoding for a simple series.

We shall discuss each of these in turn, beginning with the last. A plausible (but slightly wasteful) fixed-word representation of an integer using n bits might look like Figure 5.5, where each b_i is either 0 or 1.

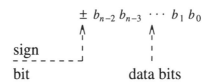

Figure 5.5.   A naive n-bit representation of an integer.

Since a sign can denote one of "+" or "−", we can devote a single bit b_{n-1} to represent the sign (with, say, the values 0 and 1 denoting "+" and "−", respectively). To summarise, in this representation, one bit determines the sign of the integer, and the remaining *data bits* determine its value, base two. A 1 in data-bit position i means that 2^i forms part of that integer. In particular, the value of the integer Figure 5.5 is precisely

$$\pm \sum_{i=0}^{n-2} b_i\, 2^i.$$

To be completely precise, we can write the meaning of the representation as

$$(-1)^{b_{n-1}} \sum_{i=0}^{n-2} b_i \, 2^i.$$

Therefore, the quantity has a finite *series* representation. The maximum absolute value we can represent is one in which all of the b_i are 1. In our representation, therefore, the largest value we can encode is $2^{n-1} - 1$.

Exercise 5.1. Why is the above representation for integers wasteful?

Example 5.1. Suppose we devote $n = 5$ bits to our naive representation. One bit is required for the sign, leaving $n - 1 = 4$ bits for the data. The largest absolute value that we can therefore represent is

$$1111_2 \;=\; 2^4 - 1 \;=\; 15 \;=\; 2^{n-1} - 1.$$

Exercise 5.2. If the digits of a number x in base d are $d_{n-2} d_{n-3} \cdots d_1 d_0$, then we define a *right shift* of x as $0 d_{n-2} d_{n-3} \cdots d_2 d_1$, and a *left shift* as $d_{n-3} d_{n-4} \cdots d_0 0$. E.g., when $n = 5$, a right shift of 0110_2 is 0011_2, while a left shift is 1100_2. What arithmetic operations can these shifts represent?

The choice of base two happens to be convenient for a computer, as is the choice of any fixed grouping of a set of bits. For example, two-bit groupings can be used to express quantities in base four, three bits for base eight (or octal), four bits for base 16 (or hexadecimal). The fact that a computer may use different bases should not come as a surprise to us. After all, in everyday life we often use bases other than our decimal system. We may order donuts by the dozen if we are unconcerned about our waistlines, count cards by pairs, measure lengths in inches and feet, weigh objects using pounds and ounces, and measure time in years, days, hours, minutes, and seconds. In each case, we have chosen (perhaps irrationally) a non-decimal base with which to work. One important observation about our representation is that if the number to be represented lies within range, then it can be represented exactly. We stress again that the choice of base has everything to do with the mechanics of computers and little to do with human concerns. Our discussion of base-two representations does account for the largest representable numbers on typical machines. A typical "short integer" representation uses 16 bits, giving rise to a maximum value using the naive representation of

$$2^{15} - 1 \;=\; 32\,767.$$

Exercise 5.3. A typical "long int" representation consists of 32 bits. What is the maximum long int value expressible using the naive representation?

The final observation made earlier is now also apparent in our representation. If A and B are two representable integers, then an arithmetic operation on them is not necessarily representable using the same number of bits. For example, if A is the maximum representable value $2^n - 1$ and B is 1, then $A + B$ is not representable. In this case, we say that an *overflow* has occurred. On some machines, this occurrence can be detected and a user-specified routine is automatically called. It is also possible for us to write code to check this ourselves, although this would potentially mean that almost every arithmetic operation would have to be preceded by an overflow check.

By way of contrast, a symbolic or a *variable-length* representation would in principle encounter none of these problems. If greater precision is required, a variable-length representation would simply allocate more space for the quantity. This is also the case for a symbolic system. On the other hand, unless great care is taken, such a representation is likely to run more slowly on a real computer.

5.1.1.2. Fixed-Point Numbers

In many applications, a fixed fractional precision is often all that is needed. For example, a weigh scale at a cheese shop is rarely accurate below about 0.001 kilogram, and most customers are satisfied with this level of precision. Similarly, we are content to balance our cheque-books to $0.01 accuracy. Just as we may devote a fixed number of bits to the representation of an integer, we may similarly set aside a fixed number of bits to represent a fraction. This is called a *fixed-point* representation, and once again, many such representations are possible.

The most obvious fixed-point representation is simply to assign one or more extra bytes that encode the fraction. Our line-drawing example in Chapter 1 did essentially this. Very few high-level programming languages support fixed-point numbers. COBOL is a notable exception.[2] More commonly, fixed-point representations are defined on-the-fly by programmers for a specific application. In this case, it is common to represent fixed-point numbers by setting aside the rightmost k bits of an integer and performing appropriate shift operations to implement the necessary arithmetic operations for the application. The number of bits devoted to a fractional representation is called its *precision*.

Exercise 5.4. If a and b are two fractional quantities each containing k decimal digits, how many decimal digits are required to hold the exact result of $a \times b$? Does the same result hold for representations in other bases, such as base two?

It is clear from this exercise that exactly representing the result of a sequence of simple arithmetic operations requires ever-increasing precision. In a fixed-point representation, this is clearly not possible, and we must therefore fit the result of an arithmetic operation into the prespecified precision available. We have two choices available to us. Suppose we have a decimal representation allowing only three digits of fraction. Then we can form a three-digit fraction

---

2. Most COBOL compilers actually implement fixed-point numbers as *floating-point numbers*, but they do ensure that input and output of them is consistent with the fixed-point format specified by the user.

from the number 1.3247 either by *truncating* or *chopping* the result to 1.324, or by *rounding* the result to 1.325. Sometimes rather obscure rules for rounding are imposed when the last digit is one-half the base (in our case 10/2=5). For example, 1.3245 may in some cases be rounded *down* to 1.324, while 1.3255 would be rounded *up* to 1.326. The rule being employed here is that when the first digit beyond our precision is one-half the base of the representation, then rounding should occur in the direction that takes the last fractional digit to an even number, as in the above examples. This is called *proper rounding*. It may not be obvious why this technique is used. Consider the following argument. If we consistently round *up*, then the actual error we make is always positive. When using proper rounding, our error is positive half the time, and negative the other half. After many rounding operations, we would expect (all other things being equal, which of course is unrealistic) our mean and overall error to both be zero.

5.1.2. The Representation of Floating-Point Numbers

5.1.2.1. Basic Definition

For many applications, neither the use of fixed-point nor variable-length representations is practical. On one hand, the amount of fixed-point precision required may not be predictable at the outset. On the other hand, a symbolic variable-length representation may be too storage intensive to be useful, particularly if only a modest amount of precision is ultimately needed. If we think of a computer representation of a number as a sequence of digits, a *floating-point* representation is one in which, figuratively, the decimal point "floats," that is, it can appear at various positions in the sequence of digits. In a fixed-point representation, of course, the point at which the fraction begins is locked in place.

A floating-point representation does *not* represent all real numbers. It has no hope of even representing a smaller set, namely the rationals, since a fixed-precision floating-point representation actually denotes a finite set of values. Floating-point representations are derived from so-called *scientific notation*, in which a number is expressed in terms of a numeric value together with an exponent in a given base. For example,

$$1.64 \times 10^4, \qquad -0.0345 \times 10^{-74}, \qquad 840.95 \times 10^{16}$$

are all numbers in scientific notation. In this formulation, the exponent is assumed to be a (signed) integer, and the numeric value can be any real number. In a floating-point representation, we impose the following rules on its form:

- The exponent is an integer of fixed size.
- The numeric component, often called the *mantissa* or *significand*, is a signed fraction of a fixed precision and a fixed base—usually 2, 8, or 16.
- In a *normalised* floating-point representation, the leading (leftmost) digit of a mantissa is non-zero unless the number being represented is zero.

Figure 5.6. A schema for floating-point number representation.

According to these rules, a floating-point number is of the form given schematically in Figure 5.6. In this case, the mantissa is a fraction which we assume to have no leading zeros, and the exponent is taken with respect to a given base. In a decimal representation, for example, we might allow five digits for a fraction and two digits for an exponent. In a binary implementation, the sign of the exponent is typically represented implicitly using *excess-n notation*. The main advantage of this approach is that an exponent of zero has only one representation (not ± 0). The idea behind offset notation is to number our exponent values from zero, which would represent the largest allowable negative exponent.

Example 5.2. The Kelvin temperature scale is such that 0 denotes about −273 degrees Celsius. Thus the Kelvin scale is an excess-273 representation for the Celsius scale.

Suppose, for example, we choose to have an eight-bit exponent. This gives us $2^8 = 256$ distinct values for an exponent. A reasonable range for exponents would be, say, $-127, -126, \cdots, 0, 1 \cdots, 128$. We then let 0 correspond to exponent value −127, 1 to −126, and so on up to 255 for exponent value 128. This is called *excess-127 notation* and is strictly for non-human consumption.

Typical floating-point representations come in two flavours: *single-precision* and *double-precision*. Normally, a single-precision number is a 32-bit quantity, with the mantissa being 24 bits including a sign, and the exponent being eight bits with a sign. A double-precision number usually occupies 64 bits, with about a 53-bit mantissa and an 11-bit exponent. The numbers can vary slightly from this on different architectures.

As an illustration, suppose we have a floating-point representation consisting of a five decimal-digit mantissa and a two decimal-digit exponent. This would allow normalised floating-point numbers such as

$$0.16400 \times 10^5, \qquad -0.34500 \times 10^{-75}, \qquad 0.84095 \times 10^{19}$$

which are floating-point representations for the numbers in scientific notation above. If we were to try to represent

$$840.956 \times 10^{16},$$

however, we would have to resort to rounding as was considered earlier with fixed-point representations.

Exercise 5.5. What is the rounded representation of 840.956×10^{16} using a four-digit decimal mantissa? How about a five-digit mantissa?

The standard arithmetic operations can be defined on floating-point numbers, but because our representation is of finite precision, a variety of numerical errors and *exceptions* can arise:

- Every operation can cause *roundoff* errors: in five-digit precision, for example,

$$0.11114 \times 10^0 \ + \ 0.53 \times 10^{-4} \ = \ 0.11114 \times 10^0 \ + \ 0.000053 \times 10^0$$
$$= \ 0.111193 \times 10^0$$
$$\approx \ 0.11119 \times 10^0.$$

We have therefore incurred a roundoff error of 0.3×10^{-5} for simple addition. This example also illustrates that normalisation is not an unwavering rule. In order to perform an operation, it is sometimes helpful to "denormalise" one number with respect to another, perform the operation, and then renormalise.

- If a number is inherently of higher precision, then we must accept a certain degree of *representation error*. This quantity is analogous to roundoff error and is defined as follows. If x is a real number and $\hat{x}$ is its representation in a floating-point system, then

$$\rho_x \ = \ |x - \hat{x}|.$$

- An *overflow* exception arises if the exponent resulting from an arithmetic operation cannot be represented. For example, in our two decimal-digit exponent,

$$0.2 \times 10^{63} \ \times \ 0.7 \times 10^{45} \ = \ 0.14 \times 10^{108},$$

which cannot be represented in the two digits allotted for the exponent.

- An *underflow* exception arises when the required exponent is smaller than the smallest possible exponent. In other words, the number is too close to zero without being zero. Thus

$$0.2 \times 10^{-63} \ \times \ 0.7 \times 10^{-45} \ = \ 0.14 \times 10^{-108}$$

would cause an underflow.

Other exceptions such as division by zero can also be identified, and handling them depends very much on the machine on which the numerical program is running. Furthermore, the actual representation of floating-point numbers and the mechanics of operating on them also vary from computer to computer. This is unfortunate, since this means that the behaviour of our programs will not be completely predictable when run on different computers. However, the situation is not entirely grim, as we shall soon discuss.

Exercise 5.6. Identify the conditions under which, for floating-point number x

$$\left(\frac{1}{x}\right) \times x = 1,$$

in a general floating-point representation consisting of a k-digit mantissa and e-digit exponent, all in base b. Start with $k = 5$, $e = 2$, and $b = 10$.

Exercise 5.7. Let us assume that floating-point operations such as addition, multiplication, subtraction and division are performed exactly up to the allotted precision. Why is it that floating-point multiplication and addition are commutative operations? An operation $\oplus$ is *commutative* if $a \oplus b = b \oplus a$.

Exercise 5.8. Show that floating-point multiplication neither *distributes* over addition, nor is it *associative*. That is, show that for suitably chosen a, b, c,

$$a(b + c) \neq ab + ac.$$

Furthermore, show that for (another set of) suitably chosen a, b, c,

$$a(bc) \neq (ab)c.$$

5.1.2.2. Specific Floating-Point Representations

We indicated earlier that the behaviour of numerical programs may differ among different computers because the implementation of floating-point representations, operations, and attendant exceptions may be quite different on each machine. Certain trends have prevented this situation from being completely chaotic.

First, on most machines, floating-point representations come in two varieties that typically differ only a small amount from machine to machine. While it is true that some computations are extremely sensitive to specific characteristics of a floating-point representation, the bulk of our computations are not. The notion of 32-bit single- and 64-bit double-precision numbers has been a popular design choice for quite some time. A great deal of special-purpose floating-point hardware has been constructed with these specific representations in mind. There are compilers and some hardware that support *quadruple-precision* numbers, and some programming languages and libraries even support *variable* or *multiple-precision* representations; in this case the precision can be specified by the user or can be modified on the sequence of floating-point operations to be performed.

Second, in the 1980s, manufacturers, engineers, and numerical analysts managed to converge to a *floating-point standard*. Sponsored by IEEE, this standard has become widely implemented. It defines the data format of floating-point numbers, the semantics of operations on them, and the semantics of the exceptions that can arise in the process of performing arithmetic operations. The data format for single- and double-precision numbers in the IEEE standard are as in Figure 5.7. The interpretation of these is slightly complicated, to allow for the special encoding of important values. It is worth discussing this to give a sense of how simple things become complex once every possible piece of information is squeezed out of every bit in the representation.

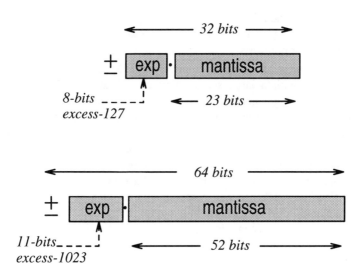

Figure 5.7.   IEEE standard representation for single precision (above) and double precision floating-point numbers.

Two complications are that:

- It is assumed that the mantissa is always normalised. In binary, this means that the leftmost bit of the fraction must be 1, unless the number is zero. If we develop a special representation for zero, then there would be no need to store this leftmost bit of information.

- Special representations for zero and usually for infinity are used. In the single-precision representation, for example, the representable exponent range would normally run from -127 to $+128$, or in excess-127 notation, from 0 to 255. In the IEEE standard, the first and last exponent values of 0 and 255 are reserved for special numbers such as 0 and ∞. Thus the actual exponent range runs from -126 to $+127$ in base two. This roughly corresponds to $10^{\pm 38}$. The double-precision format has the same rules regarding special values, giving rise to a remaining exponent range of $-1\,022$ to $1\,023$, which is about $10^{\pm 308}$.

The standard also stipulates the required accuracy of the standard arithmetic operations. Lastly, the standard indicates the exceptions that can arise when arithmetic operations are performed on floating-point numbers. These exceptions are such that they can be trapped by user-written software, and have a standard reporting format. We shall not go into details of this here.

It is important to note that in general a standard defines *minimum* performance requirements and typically does so without defining an implementation strategy.

In this case, the "standard" also recommends partial implementation strategies, in that it suggests using higher precision in some cases, even if the final result is to be reported in lower precision.

The manufacturers of hardware typically bundle the minimal floating-point standard together with a large set of auxiliary operations. The result is then fabricated onto a single computer chip that can act as a coprocessor with a computer's central processing unit. Since the arithmetic computation speed of the coprocessor is often faster than that of the CPU, some rather non-intuitive consequences can arise. For example, on some machines, (single-precision) floating-point arithmetic is *faster* than integer arithmetic. Furthermore, many functions that are time consuming to compute have been built into hardware. In this case, it can be much faster to compute $\sin x$ or $\cos x$ or even their inverses, than to evaluate a point on a polynomial curve. In general, however, it is fairly safe to assume that integer operations are usually faster than their floating-point counterparts. Moreover, the computation of transcendental functions is usually slower than the computation of polynomials.

5.2. Polynomial Series

Apart from the fact that it is important to have a familiarity with floating-point number representations, another useful reason for studying them in this chapter is that they give us a tangible idea of the amount of representation error incurred when we encode a real number requiring arbitrary precision into a floating-point number. Recall that if x is a real number and $\hat{x}$ is its representation in a floating-point system, then the representation error is

$$\rho = |x - \hat{x}|.$$

If, for example, we maintain a five decimal-digit mantissa, and we perform rounding up, then for values x whose exponent part is zero,

$$\rho \le 0.5 \times 10^{-5}.$$

In a representation having d digits of precision, our representation error is no more than 0.5×10^{-d}. We shall similarly see that when we approximate functions using some kinds of series, we can often derive a bound for the error by looking at the next term in the series.

5.2.1. Taylor Polynomials

The use of polynomials for approximation and interpolation has been a theme of this book, and the use of Taylor polynomials and series is consonant with this theme. Suppose our aim is to model the behaviour of an arbitrary function f near a point $a \in \mathbf{R}$ using a polynomial p of some degree. As we add independent constraints, we saw earlier that the degree of the polynomial satisfying these constraints must necessarily rise. A plausible set of constraints would be that, assuming n derivatives of f exist,

$$p(a) = f(a), \quad p'(a) = f'(a), \quad p''(a) = f''(a), \quad \cdots, \quad p^{(n)}(a) = f^{(n)}(a),$$

where the notation $f^{(n)}$ denotes the nth derivative of f. This says that at a, p interpolates f, and likewise each of the n derivatives of p interpolates the corresponding derivative of f. This is $n+1$ independent constraints, so we would expect p to be of degree n. It is easy to verify that the solution to this set of constraints must be

$$p(x) \;=\; f(a) + f'(a)(x-a) + \frac{1}{2!}f''(a)(x-a)^2 + \cdots + \frac{1}{n!}f^{(n)}(a)(x-a)^n.$$

The polynomial p is called the nth-degree Taylor polynomial. While p is engineered to behave like f very close to a, there is no guarantee that p will continue to do so as we move away from a.

Exercise 5.9. Verify symbolically using Maple or your own wits that the above constraints are satisfied by p.

Exercise 5.10. If f is a degree-n polynomial, verify that p is exactly f.

Maple provides a wonderful playground for gaining familiarity with Taylor polynomials and series. It always pays to work out and plot a few simple Taylor polynomials by hand. Consider the function $y = e^x$. This is the classic example with which to begin since the computation of e^x is so important, and because for any n, the nth derivative of e^x is itself. Since this function is 1 at $x = 0$, an nth-degree Taylor polynomial p_n about 0 that approximates e^x is

$$p_n(x) \;=\; 1 + x + \frac{1}{2}x^2 + \cdots + \frac{1}{n!}x^n.$$

Figure 5.8 plots p_2 against e^x. It is clear that even a quadratic does quite well in a short interval about the origin. As we add terms to the polynomial, we expect the approximation to be a good one over a wider interval. It is also instructive to form the *residual error* term

$$r_n(x) \;=\; f(x) - p_n(x),$$

to study the error behaviour of the approximation. Figure 5.9 plots r_2 and r_4. The more flat the curve stays at zero, the better the approximation. In this case, r_4 does very well on the interval $[-1,+1]$. While p does a good job of approximation, as it stands its form is not amenable to practical computation. The factorial terms grow extremely quickly which can lead to inaccuracies in floating-point computation; if high precision is required, the terms in the series must be computed more carefully. We shall return to this problem shortly.

We have so far seen that a Taylor polynomial can do a good job of approximating a function over some region about the expansion. It is therefore natural to extend the notion of a Taylor polynomial of finite degree n to that of a

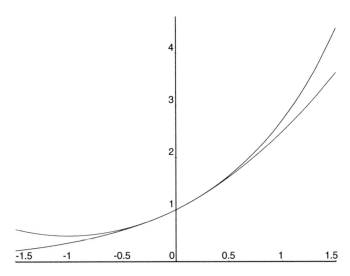

Figure 5.8.   The function $y = e^x$ and a quadratic Taylor polynomial approximation.

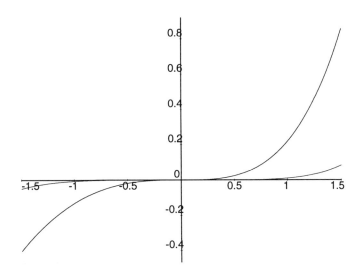

Figure 5.9.   Residual errors $r_2(x)$ and $r_4(x)$.

series by letting n grow arbitrarily large. In this case, we have defined a *Taylor series* that, if it converges, should be a good approximation to f:

$$f(x) = \sum_{n=0}^{\infty} \frac{1}{n!} f^{(n)}(a)(x-a)^n.$$

The resulting series is called the *Taylor series* (or *expansion*) of f about a. The simple fact that we can write down such an expression does not in any way guarantee that the series actually converges. Moreover, there are functions for which the series converges, but not to the desired function. These topics are properly in the domain of a mathematical analysis course, but we must be aware of the fact that while Taylor series is an extremely useful tool, we must take into account the bounds of convergence, if any. The fact is that a large class of functions have converging Taylor series approximations.

The series representation also tells us the error we make when we *truncate* the series after the nth derivative. This of course is exactly the notion of residual error defined above, since truncating the series after n derivatives gives rise to a Taylor polynomial of degree n. There is a useful way of placing a bound on the residual error. To do so requires *Taylor's theorem*, which states that if the $n+1$ derivatives of f all exist on an interval $[x_0, x_1]$ that contains a, then there is a value $c \in [x_0, x_1]$ such that

$$p(x) = f(a) + f'(a)(x-a) + \frac{f''(a)(x-a)^2}{2!} + \cdots + \frac{f^{(n)}(a)(x-a)^n}{n!}$$
$$+ \frac{f^{(n+1)}(c)(x-a)^n}{(n+1)!}.$$

This allows us to give an upper bound for the representation error r_n:

$$r_n(x) \le C |x-a|^{n+1},$$

where C is a constant given by

$$C = \frac{|f^{(n+1)}(x_{\max})|}{(n+1)!},$$

and $x_{\max}$ is a point in $[x_0, x_1]$ for which $f^{(n+1)}(x)$ is a maximum. Another way of saying this is that r_n is of *polynomial order* $(x-a)^{n+1}$, which is typically written

$$r_n(x) = O((x-a)^{n+1}).$$

Example 5.3. The Taylor series expansion of e^x is

$$e^x = 1 + x + \frac{x^2}{2!} + \frac{x^2}{3!} + \cdots .$$

It can be shown that this sequence converges for all $x \in \mathbf{R}$. The analysis of a Taylor series for e^x is a standard problem in numerical analysis courses because, despite its apparent simplicity, it illustrates several numerical problems. We will examine some of these problems in the next few exercises and examples. A much broader treatment of this can be found in many numerical analysis textbooks.[3]

Exercise 5.11. Using Taylor's theorem, what is the residual $r_n(x)$ for e^x?

Exercise 5.12. The following fragment of C code computes an n-term Taylor's series approximation to e^x.

```
double Exp(x,n)
    double x;
    int n;
{
    long int fact();        /* fact(n) means n! */
    int i;
    double s, power();      /* power(x,i) means x^i */
    s = 1.0 + x;
    for (i=2; i<=n; i++)
        s += 1.0/fact(i)*power(x,i);
    return(s);
}
```

For $n = 10$, we get about two decimal places of accuracy for $x \in [-3, +3]$. For n even a little larger than this, the routine misbehaves. Execute this code and explain its behaviour. You will have to implement **fact()** and **power()** as well. Do so efficiently.

Exercise 5.13. A much more stable version of **Exp** can be derived by changing its main loop into a Horner's rule decomposition of the polynomial *coefficients* so that the factorial terms are never computed by themselves. Rewrite that loop and compare its accuracy to the original **Exp**.

Exercise 5.14. While Taylor's theorem provides us with an upper bound for the residual error r_n, it is sometimes too coarse to be of use to analyse specific behaviour, depending on the assumptions used. A practical measure of the quality of error is to compute as many terms in the series as are needed before the addition of the next term makes no difference to the result. Assuming no numerical instability arises in the process, this gives us the best approximation for e^x that is possible with the precision we have. Modify your routine from Exercise 5.13 to perform this test. Now the parameter **n** is no longer needed, but you might want to keep it to indicate the *maximum* number of terms for which the user is willing to pay. Compute the number of terms required to achieve full double-precision accuracy for values of x on the interval $[-15, +15]$. Write a new version of your code with every reference to **double** replaced by **float**, which is C's version of a single-precision floating-point data type. Perform the same test. Comment on the results.

Exercise 5.15. For the benefit of those who have not done the previous exercise, Figure 5.10 depicts the number of terms in the series required to achieve full double precision.

3. See, for example, J.L. Morris, *Computational Methods in Elementary Numerical Analysis*, John Wiley and Sons, New York, 1983; or R.L. Johnston, *Numerical Methods, A Software Approach*, John Wiley and Sons, New York, 1982.

Two rather interesting observations can be made. First, the number of terms required increases with the magnitude of x. Second, we need more terms for negative values than for positive values. Explain why this is the case. Hint: construct a sequence of partial results to see how each term in the series contributes to the overall result.

Exercise 5.16. Use properties of exponentiation to modify your code so that the same number of terms is required for both **Exp(x)** and **Exp(-x)**.

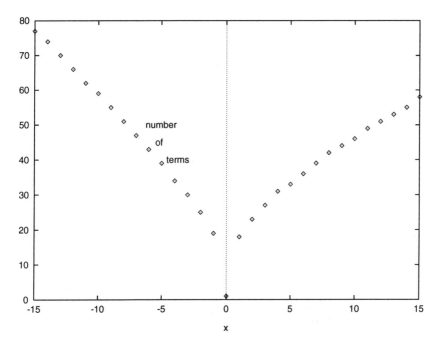

Figure 5.10. The number of required terms in the series plotted against x.

Perhaps the most useful aspect of a Taylor series is that it provides a foundation upon which to perform error analysis. For example, in the last chapter, we discussed various integration techniques, but did not discuss their error. We are now equipped to do so.

5.2.2. Error Analysis of Quadrature Algorithms

5.2.2.1. Error on a Single Subinterval

The derivations in this section are very simple but are rather notationally heavy. The reader is invited to step through this analysis as the author did by using Maple. The first step will be to describe the expected error of Riemann sums, the

trapezoid rule, and Simpson's rule over a single subinterval, which we shall label
$[a,b]$. We shall subsequently look at the corresponding compound rules.

We begin with an analysis of Riemann sums. In this case, on an arbitrary
subinterval $[a,b]$, we compute the area of a rectangle whose height is given by
evaluating f, the function to be integrated, at the midpoint of the interval. The
height is given by $f(m)$, where $m = (a+b)/2$. Let us compute a Taylor series
expansion of f about m:

$$f(x) = \sum_{n=0}^{\infty} \frac{1}{n!} f^{(n)}(m)(x-m)^n$$

$$= f(m) + f'(m)(x-m) + \frac{f''(m)(x-m)^2}{2} \tag{1}$$

$$+ \frac{f'''(m)(x-m)^3}{6} + \frac{f^{(4)}(m)(x-m)^4}{24} + \cdots.$$

This is an expression for a "curve" going through the midpoint of the interval.
We integrate this expression term-by-term over $[a,b]$. Some of the terms
disappear, giving us the following result.

Error on a Single Interval Using Riemann Sums.

$$\int_a^b f = f(m)(b-a) + \frac{f''(m)(b-a)^3}{24} + \frac{f^{(4)}(m)(b-a)^5}{1920} + \cdots. \tag{2}$$

The first term is the expression for the approximation as given by the Riemann
sum on $[a,b]$, and the residual is of order Δx^3, where $\Delta x = b-a$.

This notion of *order* is as above, except that we now have a definite value, namely
b, for the indeterminate x. The interpretation is that we expect the error made to
vary with the cube of the length of our interval $[a,b]$.

Perhaps at this point the reader can see the way the game is played. Our goal
is to derive an expression for the definite integral whose first term coincides with
the quadrature method we are attempting to analyse. The remaining terms thus
give the error of the technique.

We shall derive an expression for the trapezoid rule as follows. We first
return to our Taylor series expansion, Eq. 1, and rewrite it so that we have an
expression for $f(m)$:

$$f(m) = f(x) - f'(m)(x-m) - \frac{f''(m)(x-m)^2}{2} \tag{3}$$

$$- \frac{f'''(m)(x-m)^3}{6} - \frac{f^{(4)}(m)(x-m)^4}{24} + \cdots.$$

By letting $x = a$ and $x = b$, we get two expressions for $f(m)$. Add these two expressions together, divide by two, and after much simplification we arrive at

$$f(m) = \frac{f(a) + f(b)}{2} - \frac{f''(m)(b-a)^2}{8} - \frac{f^{(4)}(m)(b-a)^4}{384} - \cdots . \quad (4)$$

We could integrate expression termwise as before. Another common approach is to take this expression for $f(m)$ and substitute it into our expression for Riemann sums (Eq. 2). This gives (again after much simplification) the error using the trapezoid rule.

Error on a Single Interval Using the Trapezoid Rule.

$$\int_a^b f = (b-a)\frac{f(a) + f(b)}{2} - \frac{f''(m)(b-a)^3}{12} - \frac{f^{(4)}(m)(b-a)^5}{480} - \cdots . \quad (5)$$

The first term is the expression for the trapezoid rule. As we can see, the order of error is the same as for Riemann sums. This confirms our observation in the previous chapter that roughly the same number of function evaluations was required to compute the definite integral to the same precision. The interested reader is invited to consult the appendix at the end of this chapter which presents the Maple code used to compute the previous two series.

We can synthesise a formula for Simpson's rule in an analogous way. We just find an appropriate weighting of the other rules that yields Simpson's rule as a term in the series. In this case, we add together twice the series for Riemann sums (Eq. 2) and the series for the trapezoid rule (Eq. 5). Dividing by three to normalise gives us the error for Simpson's rule.

Error on a Single Interval Using Simpson's Rule.

$$\int_a^b f = \frac{(b-a)}{6}\left(f(a) + f(b) + 4f(m)\right) - \frac{f^{(4)}(m)(b-a)^5}{2880} - \cdots . \quad (6)$$

This equation thus predicts order Δx^5 error, a considerable improvement. Thus our set of tests showing that Simpson's rule had lower error is vindicated by the theory. Assessing the error of Catmull-Rom integration is complicated by the fact that the cubic defined on the interval $[a,b]$ requires two points outside that interval for computation. Because of this technicality, we omit the error analysis.

We have just seen how a Taylor series provides a springboard to establishing rigorous error bounds on the behaviour of quadrature algorithms. We now briefly consider the error made by piecewise or compound approaches.

5.2.2.2. Error in Piecewise or Compound Quadrature

Since each piece in our piecewise approaches to integration operate over distinct portions of the curve, the error incurred over each panel is independent of the other. It is thus appropriate to define the overall error as the sum of the errors made over individual pieces. Since we have already discussed the single-interval case, we have an immediate error estimate for the compound cases. If the width of the subintervals is the same, then we can simplify our expressions.

We first consider the trapezoid rule. The analysis for Riemann sums is identical. If the spacing between knots is uniformly Δx, then the above analysis states that the error over a single subinterval $[x_i, x_{i+1}]$ is

$$\int_{x_i}^{x_{i+1}} f = \Delta x \, \frac{f(x_i) + f(x_{i+1})}{2} - \frac{f''(m_i) \Delta x^3}{12} + \cdots,$$

where $m_i = (x_i + x_{i+1})/2$. If as before our domain of integration $[a,b]$ is subdivided into n subintervals $I_i = [x_i, x_{i+1}]$ then our overall error is

$$\int_a^b f = \sum_{i=0}^{n-1} \int_{x_i}^{x_{i+1}} f \approx \Delta x \left(\frac{f(a)}{2} + \sum_{i=1}^{n-1} f(x_i) + \frac{f(b)}{2} \right) - \frac{\Delta x^3}{12} \sum_{i=0}^{n-1} f''(m_i).$$

In practice, computing n second derivatives to estimate the residual error is too expensive. Recall from our discussion of Taylor's theorem above that a suitable x_{max} exists that would bound the error in f''. That is, $f''(m_i) \le f''(x_{max})$ for all subintervals I_i. This would imply that

$$\frac{\Delta x^3}{12} \sum_{i=0}^{n-1} f''(m_i) \le \frac{\Delta x^3}{12} n f''(x_{max}).$$

Finding x_{max} is usually not at all easy, so instead it is usually assumed that f'' is sufficiently smooth so that any $x \in [a,b]$ will do the job. The numerical analysts like to denote this choice of x by a fancy Greek letter ξ (called xi). Sometimes, ξ is actually used to represent x_{max}. For the purposes of our analysis, this is unimportant. Our overall integral with residual error term now looks like

$$\int_a^b f = \Delta x \left(\frac{f(a)}{2} + \sum_{i=1}^{n-1} f(x_i) + \frac{f(b)}{2} \right) - \frac{\Delta x^3}{12} n f''(\xi).$$

Now note that $n \Delta x = b - a$. This finally gives us our error bound using the compound trapezoid rule.

Error Bound Using the Compound Trapezoid Rule on Equally-Spaced Knots.

$$\int_a^b f = \Delta x \left(\frac{f(a)}{2} + \sum_{i=1}^{n-1} f(x_i) + \frac{f(b)}{2} \right) - (b-a)\frac{\Delta x^2}{12} f''(\xi).$$

The residual error is a different quantity than that defined earlier. We shall denote the residual error in a compound quadrature scheme on n subintervals (or $n+1$ points) by R_n. In the case of the trapezoid rule,

$$R_n = O(\Delta x^2).$$

The argument for Simpson's rule is virtually identical to the above. After tediously grouping together terms and assuming that $f^{(5)}$ is smooth, we arrive at the following rule.

Error Bound Using Compound Simpson's Rule on Equally-Spaced Knots.

$$\int_a^b f = \frac{\Delta x}{3} \left(f(a) + 4f(x_1) + 2f(x_2) + \cdots + 4f(x_{2n-1}) + f(b) \right)$$

$$- (b-a)\frac{\Delta x^4}{180} f^{(4)}(\xi).$$

It is easiest to make sense of these error formulae by considering what is gained if the sample spacing is halved. For example, suppose we employ the compound trapezoid rule using $2n$ intervals instead of n intervals. We can then look at the relationship between the residual errors R_n and R_{2n}:

$$|R_n| = \frac{b-a}{12} \Delta x^2 \, |f''(\xi_n)|.$$

$$|R_{2n}| = \frac{b-a}{12} \left(\frac{\Delta x}{2} \right)^2 |f''(\xi_{2n})|.$$

We have assumed that the errors R_n and R_{2n} are bounded above by maximal errors at ξ_n and ξ_{2n} respectively. These two values need not be the same, but if we assume that indeed f'' is fairly smooth so that $f''(\xi_n)$ and $f''(\xi_{2n})$ are almost equal, then

$$|R_{2n}| \approx \frac{|R_n|}{4}.$$

Under these assumptions, our measure predicts that by doubling the number of intervals, and thereby halving their width, we should reduce our error by a factor of four.

Let us see if this theory is consistent with reality. Suppose we wish to perform the following integration numerically:

$$\int_0^1 \frac{4}{1+x^2}\, dx = \pi.$$

This function is actually four times the derivative of $\tan^{-1}x$. The second derivative of the integrand exists and is continuous everywhere on the interval $[0,1]$; however it is quite variable over the interval. The function together with its second derivative is plotted in Figure 5.11.

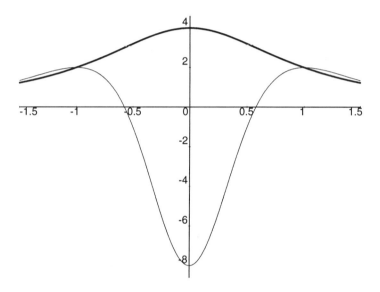

Figure 5.11. A plot of $f(x) = 4/(1+x^2)$ (thick curve) together with its second derivative.

The second derivative f'' clearly achieves a global minimum and largest absolute value at $x = 0$ so we could apply Taylor's theorem at that point to show the $O(\Delta x^2)$ convergence behaviour of the trapezoid rule. However, since we know the value of the integral, it is more interesting to compute the *true* relative error

$$\tau_n = \frac{|A_n - \pi|}{\pi},$$

where A_n is the computed integral on n subintervals, to see if τ_n/τ_{2n} has the same behaviour as that predicted for R_n/R_{2n} in our analysis. The results can be found in Table 5.1. The improvement rate on the actual error is as predicted even in this case where our assumptions about f'' are somewhat suspect. In this case, the convergence rate even increases slightly as the number of samples increases.

n	A_n	τ_n	τ_n/τ_{2n}
2	3.1000000	.013239	–
4	3.1311765	.003316	3.993080
8	3.1389885	.000829	3.999826
16	3.1409416	.000207	3.999992
32	3.1414210	.000052	4.000006
64	3.1415520	.000013	4.000492
128	3.1415825	$.324 \times 10^{-5}$	4.002066
256	3.1415902	$.808 \times 10^{-6}$	4.007095
512	3.1415921	$.191 \times 10^{-6}$	4.235392
1024	3.1415925	$.382 \times 10^{-7}$	4.991667

Table 5.1. Error estimates for f using the compound trapezoid rule.

5.3. Non-Polynomial Series: Trigonometric Fourier Series

5.3.1. Definition

In Chapter 3, we discussed the notion of the "frequency" of a function and related the minimum sampling rate to a function's maximum frequency. While this was easy to do for functions of the form $\cos nx$, $\sin nx$ for constant $n \in \mathbf{R}$, it was not at all clear what is meant by the "frequency" of an arbitrary function. However, if we can accurately approximate functions in terms of such sines and cosines, then we can make sense of this idea. This is among the many things that Fourier analysis does. The theory and application of Fourier analysis is far richer than we can even begin to describe here. It is an indispensable tool in a great many scientific and engineering applications. Nevertheless, even a little exposure to the approach yields pleasant insights.

Fourier analysis is based on the representation of arbitrary functions by the weighted sum of simpler functions. These functions are taken from a family of *orthogonal functions*. In our case, we shall be using a family of functions of the form $\{\cos mx, \sin nx : m,n \in \mathbf{Z}\}$, but other families can be used, including special sets of polynomials. For an *orthonormal* basis, these functions are normally divided by π, and the constant "function" $1/(2\pi)$ is added. Such families allow us to construct discrete *Fourier series* when a function is periodic, and a *Fourier transform* when the function is not, or when the domain over which we investigate the function is infinite.

Exercise 5.17. The *inner product* of two functions f and g on a domain $[a,b]$ is given by

$$<f,g> = \int_a^b f(x)\, g(x)\, dx.$$

This is a direct generalisation of the *dot* product for vectors (try to think of f and g as representing infinite-dimensional vectors). Just as in a dot product, two functions f, g are *orthogonal* if $<f,g> = 0$. Use Maple to show that $\{\cos mx, \sin nx : m,n \in \mathbf{Z}\}$ are an orthogonal family on $[-\pi, +\pi]$ by verifying that for integral n,m,

$$<\cos mx, \sin nx> = 0,$$

$$<\cos mx, \cos nx> = 0, \quad m \neq n,$$

$$<\sin mx, \sin nx> = 0, \quad m \neq n.$$

Furthermore, show that

$$<\sin mx, \sin nx> > 0,$$

$$<\cos mx, \cos nx> > 0, \quad m = n.$$

Let f be a function defined over an interval of length 2π. We shall usually take this interval to be $[-\pi, +\pi]$, although later we will allow it to expand. Normally, f is assumed to be *periodic*, meaning that it replicates itself over intervals of the form $[k\pi, (k+2)\pi]$, $k \in \mathbf{Z}$. More formally, f is periodic on $[-\pi, +\pi]$ if for $k \in \mathbf{Z}$ and $x \in [-\pi, +\pi]$, $f(x + 2k\pi) = f(x)$. For our purposes, we will not go beyond the primary interval of $[-\pi, +\pi]$, so we will not notice the periodicity. We define the Fourier series of interest to us in the following box.

Trigonometric Fourier Series of f on the Interval $[-\pi, +\pi]$.

$$f(x) = \frac{a_0}{2} + \sum_{n=1}^{\infty} a_n \cos nx + b_n \sin nx,$$

where the coefficients a_n, b_n are given by

$$a_n = \frac{1}{\pi} \int_{-\pi}^{+\pi} f(x) \cos nx\, dx, \quad n = 0, 1, 2, \cdots,$$

$$b_n = \frac{1}{\pi} \int_{-\pi}^{+\pi} f(x) \sin nx\, dx, \quad n = 1, 2, \cdots.$$

We stress that other kinds of Fourier series exist, but at the risk of adding to the confusion, we shall for brevity call the above the Fourier series of f. Shortly, we shall be computing several Fourier series.

Exercise 5.18. What is the Fourier series of $f(x) = \sin x$ and $f(x) = \cos x$ on $[-\pi, +\pi]$?

An obvious difference between Fourier series and Taylor series is that a Fourier series is based on integration, while a Taylor series is based on differentiation. A more significant difference is that a Taylor series is fundamentally tied to interpolation at a point. A Fourier series is strictly an approximation scheme. One main similarity is that both can be very tedious to compute; the Fourier coefficients a_n, b_n can be particularly messy to compute unless one can take advantage of identities involving $f(x) \cos nx$ and $f(x) \sin nx$ for specific functions f.

Several useful properties of integration help in the computation of Fourier series. First, note that if f is an *odd* function, meaning that $f(-x) = -f(x)$, then because $\cos nx$ is an *even* function (meaning that $\cos -nx = \cos nx$), their multiplication $f(x) \cos nx$ is an odd function on $[-\pi, +\pi]$. Therefore,

$$a_n = \frac{1}{\pi} \int_{-\pi}^{+\pi} f(x) \cos nx \, dx = 0, \quad n = 0, 1, \cdots,$$

and the Fourier series for odd functions f reduces to the following.

Fourier series for odd functions on $[-\pi, +\pi]$.

$$f(x) = \sum_{n=1}^{\infty} b_n \sin nx.$$

Moreover, the computation of each b_n is slightly simplified:

$$b_n = \frac{2}{\pi} \int_{0}^{\pi} f(x) \sin nx \, dx, \quad n = 1, 2, \cdots.$$

Similarly, if $f(x)$ is an even function, then by identical reasoning,

$$b_n = \frac{1}{\pi} \int_{-\pi}^{+\pi} f(x) \sin nx \, dx = 0, \quad n = 1, 2, \cdots.$$

In this case, the Fourier series simplifies to the following.

Fourier Series for Even Functions on $[-\pi, +\pi]$.

$$f(x) = \frac{a_0}{2} + \sum_{n=1}^{\infty} a_n \cos nx.$$

The computation of the a_n also reduces to

$$a_n = \frac{2}{\pi} \int_0^\pi f(x) \cos nx \, dx, \quad n = 0, 1, 2, \cdots .$$

If an odd or even function is translated in x, then it is unlikely to remain odd or even. Thus the choice of interval over which to perform the series decomposition can be important.

Fourier series can be shown to converge for a very wide class of functions. Many discontinuous functions can also be approximated. Before proceeding further, it is helpful to consider some examples.

Exercise 5.19. What is the Fourier series of $f(x) = \sin x \cos x$ on $[-\pi, +\pi]$?

Exercise 5.20. Use the above properties to compute the Fourier series of $f(x) = \cos^k x$ and $\sin^k x$, for specific positive, integral values of k. Use Maple to generalise this to arbitrary k. Hint: consider even and odd k.

5.3.2. Examples

It is unsurprising that Fourier series are good for approximating periodic trigonometric functions; after all, Fourier series are themselves periodic, and they are composed of trigonometric basis functions. It is quite remarkable however that trigonometric functions that wiggle so much can actually approximate objects such as straight line segments. Let us compute the Fourier series for the function

$$f(x) = x$$

on the interval $[-\pi, \pi]$. We first note that since f is an odd function, all of the a_n are zero, and thus the Fourier series will be of the form

$$f(x) = \sum_{n=1}^{\infty} b_n \sin nx.$$

Furthermore, the b_n can be written in closed form. Using integration by parts,

$$b_n = \frac{2}{\pi} \int_0^\pi x \sin nx \, dx$$

$$= \frac{2}{\pi n^2} (\sin nx - nx \cos nx) \Big|_0^\pi$$

$$= \frac{2}{\pi n^2} (\sin n\pi - n\pi \cos n\pi)$$

$$= -\frac{2n\pi}{n^2 \pi} \cos n\pi$$

$$= -\frac{2}{n}\cos n\pi.$$

Now, $\cos n\pi = -1$ when n is odd and $\cos n\pi = 1$ when n is even. Thus the above expression is equal to $2/n$ when n is odd and $-2/n$ when n is even. Therefore our Fourier series is

$$f(x) = 2\sin x - \sin 2x + \frac{2}{3}\sin 3x - \frac{2}{4}\sin 4x + \frac{2}{5}\sin 5x - \cdots .$$

Figure 5.12 plots an approximation to $f(x) = x$ using three terms of the Fourier series. Figure 5.13 depicts the result of using seven terms, and Figure 5.14 demonstrates the use of 20 terms.

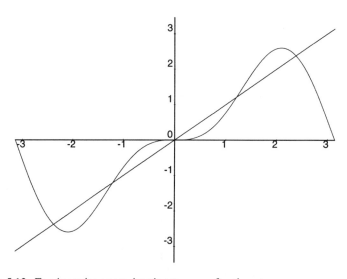

Figure 5.12. Fourier series approximation to $y = x$ after three terms.

The *Fourier spectrum* of a function f is related to the coefficients a_n and b_n in its Fourier series. Normally it is defined over the continuous versions of these coefficients arising from the *Fourier transform* of f. For our purposes, it suffices to define c_n simply as the sequence of ordered pairs:

$$c_n = (a_n, b_n).$$

We shall use this discrete sequence as an indicator of the continuous transform. In fact, the sequence of c_n becomes a continuous function under the Fourier transform, which we shall discuss later. At the outset of this book, we noted that a function f defined over the space $\mathbf{R}^n$ can be represented in many ways, including implicit, explicit, and parametric forms. All of these are called *spatial-domain*

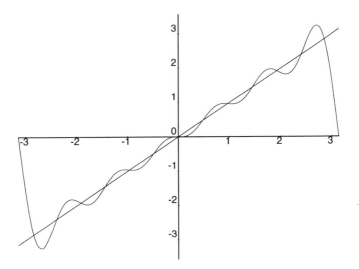

Figure 5.13. Fourier series approximation to $y = x$ after seven terms.

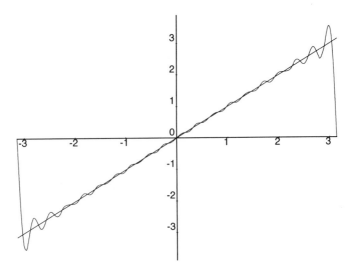

Figure 5.14. Fourier series approximation to $y = x$ after 20 terms.

representations of f. The Fourier transform, or in our case the coefficients given by the c_n, provide us with a completely different way of looking at f, namely a *frequency-domain* representation.

Let us try to make better sense of what a frequency-domain representation means by looking more closely at the discrete c_n. A *sinusoid* is a function like sine and cosine that oscillates at a fixed frequency.[4] So far, we know that rather than depicting a function in terms of its extent in x and y, a Fourier series representation causes us to think of the function as being made up of a weighted sum of sinusoids of various fundamental frequencies. Recall that the function $\sin x$ has a *(circular) frequency* of one cycle for every 2π units or $1/2\pi$ cycles per unit distance or unit time. If x refers to time, then the frequency of $\sin x$ is $1/2\pi$ cycles per second; this unit is called a *Hertz*, abbreviated Hz. If we wish to know the *angle* through which $\sin x$ travels over unit distance or time, then we multiply its circular frequency by 2π radians, since one cycle corresponds to 2π radians. This quantity is therefore measured in radians per second and is called *angular frequency*. If we just use the word "frequency," we mean circular frequency. Thus circular frequency ϕ and angular frequency ω differ by a constant factor. The frequency of $\sin 2x$ is $2/2\pi$ or $1/\pi$ Hz. In general, the frequency of $\sin nx$ for a constant real number $n > 0$ is $n/2\pi$. Note that a function's frequency is the reciprocal of its *period*, which would be $2\pi/n$. For example, the function

$$\sin \frac{1}{x}$$

has a period that approaches 0 as $x \to 0$, and therefore its frequency approaches ∞.

Exercise 5.21. What is the angular frequency of $\sin nx$ or $\cos nx$?

Exercise 5.22. What is the angular frequency of $\cos nx$, when n is a positive integer?

Exercise 5.23. What are the period and the frequency of $\sin nx$ or $\cos nx$ when $n = 0$? Discuss.

We can now make the following informal observation: the value of c_n for each n tells us how much the sinusoid with circular frequency

$$\phi_n = \frac{n}{2\pi}$$

contributes to the composition of $f(x)$. If we talk about the "frequency content" of $f(x)$, we really mean the set of sinusoids that are weighted by non-zero c_n. If we say that a function contains "high frequencies," we mean that the Fourier

---

4. While it is often convenient to think of $\sin x$ and $\cos x$ as different functions, they are of course related to one another by a phase shift of $\pi/2$ radians. For example, $\sin x = \cos(x - \pi/2)$. Any "sinusoid" can therefore be written solely in terms of $\sin x$ or $\cos x$ if desired. Indeed, a general sinusoid can be written as $\cos(\omega x + \theta)$, where ω is a frequency and θ is a *phase shift* (or offset).

spectrum of $f(x)$ has many nonzero c_n when n is large. Indeed, the ϕ_n corresponding to largest n for which c_n is nonzero is exactly the "highest frequency" to which we alluded in Chapter 3. Very few functions actually have Fourier series of finite length, so that in fact most functions contain arbitrarily high frequencies. However, a great many functions are such that $c_n \to 0$ as $n \to \infty$, and if a function f is such that its corresponding c_n tend to zero with increasing n, for all practical purposes f is bounded in frequency. A function that is uniformly zero in frequency outside of a finite interval is called *band limited*.

A fundamental theorem that we cannot prove here is that no function can be both band limited and space-time limited. A precise statement of this theorem requires a substantially more careful presentation of Fourier series.[5] We can interpret it informally with our knowledge of trigonometric Fourier series, however. If we note that the interval over which a Fourier series is taken is arbitrary, then the theorem states that if a function f is nonzero only over a finite spatial domain, then any Fourier series of f taken over an interval that includes that domain must be nonterminating; conversely if f has a terminating Fourier series, then it it has an infinite spatial domain over which it is nonzero. This fact is at the heart of a beautiful theorem called the *Heisenberg uncertainty principle*. In quantum physics it is used to describe limits of the precision with which we can model or simultaneously "observe" properties of atomic particles. For example, there is an inverse relationship between the precision with which we can *simultaneously* measure a particle's position and its momentum.[6]

Figure 5.15 depicts a continuous version of the spectrum of $f(x) = x$ which was done simply by joining the b_n by line segments. Since the a_n are zero, it suffices to look at only the graph of the b_n. In this case $n = 30$, which corresponds to a maximum circular frequency of $15/\pi \approx 4.8$. Observe that the Fourier spectrum contains arbitrarily high frequencies, but the proportion of influence steadily decreases with increasing frequency. This is more visible when we plot the magnitude of the b_n in Figure 5.16. The shape of the curve qualitatively describes the rate of convergence of the Fourier series.

Exercise 5.24. For the spectrum of the line, derive an expression for the ratio

$$\frac{|b_{n+1}|}{|b_n|}.$$

What does this value suggest about the convergence of the Fourier series (i.e., as n gets large)?

---

5. This would not be too difficult to do. We would need to discuss Fourier series in the more general context of orthogonal basis functions and inner product spaces.

6. For further reading, see A.P. French and E.F. Taylor, *An Introduction to Quantum Physics*, Norton, New York, 1978; and J.E. Marsden, *Elementary Classical Analysis*, W.H. Freeman and Co., San Francisco, 1974.

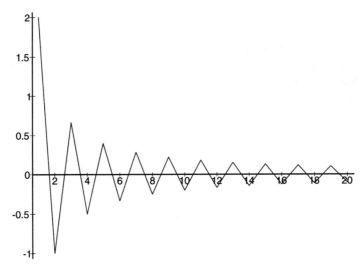

Figure 5.15. Approximate Fourier spectrum of the line $y = x$. The horizontal axis is the frequency in cycles/radian, and the vertical axis is the value of the coefficient at that frequency.

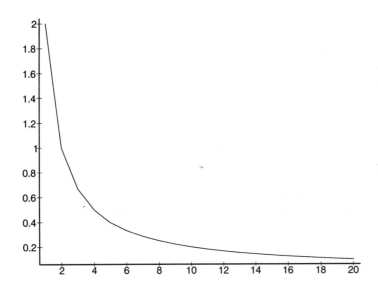

Figure 5.16. Magnitude of the Fourier spectrum of the line $y = x$.

As another example, consider the function

$$f(x) = \frac{\sin 10x}{10x},$$

which is the sinc function introduced in Chapter 3. Sinc is an even function on $[-\pi,\pi]$, meaning that there are no b_n terms. Maple was used to compute the a_n numerically. The plot of only nine terms of the Fourier series is plotted against the sinc function in Figure 5.17.

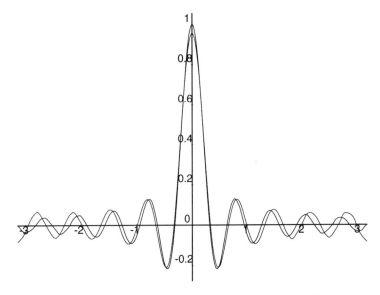

Figure 5.17. The function $\sin 10x / 10x$ plotted against a nine-term Fourier series approximation.

It is very interesting to look at the Fourier spectrum of the sinc function. After 20 terms, the function as in Figure 5.18. Although the spectrum is highly approximate, it is clear what it converges to: a box of width 10 radians and height 0.1. This link between a sinc in spatial domain and a box in frequency domain has deep implications for filtering, especially, as we shall see, since the link also works in the other direction. We shall develop these implications further after our discussion of the Fourier transform.

As a final example, let f be a *box* function defined on $[-\pi,\pi]$ as follows

$$f(x) = \begin{cases} 0 & \text{if } |x| > 1. \\ 1 & \text{if } |x| \leq 1. \end{cases}$$

This function contains two discontinuities. We have arranged for this function to

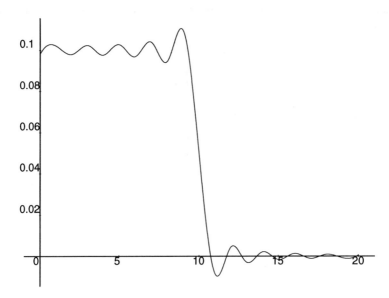

Figure 5.18. The approximate Fourier spectrum of $\sin 10x / 10x$.

be an even function, so that its Fourier series is of the form

$$f(x) = \frac{a_0}{2} + \sum_{n=1}^{\infty} a_n \cos nx.$$

Computing the a_n is easy to do by hand if we simply observe that f is nonzero only over $[-1,+1]$. Thus

$$a_n = \frac{1}{\pi} \int_{-1}^{+1} f(x) \cos nx, \quad n = 0, 1, 2, \cdots$$

$$= \frac{1}{\pi} \int_{-1}^{+1} \cos nx$$

$$= \frac{\sin nx}{n\pi} \Big|_{-1}^{+1}$$

$$= \frac{2}{n\pi} \sin n.$$

Recall that $\sin n / n \to 1$ as $n \to 0$. Our Fourier series is therefore

$$f(x) = \frac{1}{\pi} \left(1 + \sum_{n=1}^{\infty} \frac{2}{n} \sin n \cos nx \right).$$

Figure 5.19 depicts the four-term Fourier series approximation to the box function. Figures 5.20 and 5.21 depicts 20-term and 50-term approximations, respectively.

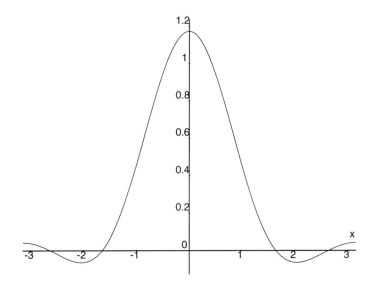

Figure 5.19. The four-term Fourier Series approximation to a box function.

These approximations illustrate some interesting properties. First, notice the *ringing* that is strongly evident near the sharp rise and fall of the box function. This behaviour is quite familiar to us from Chapter 3. While it is clear that the series will ultimately converge to the box, the ringing effect near the transition will always be present for any finite series, although the ringing gets squeezed into an ever-narrowing region. This effect is called the *Gibbs' phenomenon*. Second, the function passes *midway* between the discontinuity. That is, in the Fourier series approximation $\bar{f}$ of f,

$$\bar{f}(-1) \;=\; \bar{f}(1) \;=\; 0.5.$$

Exercise 5.25. By plotting the a_n, show that the frequency-domain representation of the box function is a sinc. Thus we see that the box function and the sinc function transform into one another.

Exercise 5.26. When appropriately normalised, the box function can be used as a continuous version of a moving-average filter (recall Chapter 3). Seeing that the frequency-domain representation of such a filter is a sinc function, is it really the case that the moving-average filter is a low-pass filter?

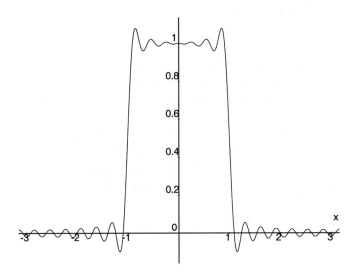

Figure 5.20. The 20-term Fourier Series approximation to a box function.

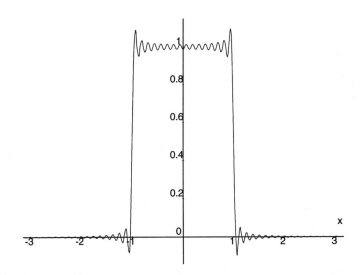

Figure 5.21. The 50-term Fourier Series approximation to a box function.

Exercise 5.27. Determine the Fourier series and estimate the frequency-domain representation of the Gaussian function

$$f(t) = \frac{1}{\sqrt{2\pi}\,\sigma} \exp\left(-\frac{x^2}{2\sigma^2}\right)$$

for $\sigma = 1$. Maple can handle the necessary integrals. Plot the Gaussian distribution for $\sigma = 1$ and its frequency domain representation as well. Comment on the results.

Exercise 5.28. In Chapter 3, we saw that the discrete Kronecker δ-function, $\delta[x]$, was useful for expressing the sampling of discrete signals. The analogous continuous version, the *Dirac δ-function*, $\delta(x)$, is an important construction that is found in many physical problems, as well as being prominent in signal theory. It is a "function" only in the most liberal sense of the term, and is in reality a remarkable example of forcing a function to have certain properties because they are so useful. Among them is the crucial *sifting property*: for any function f that is continuous at 0, and for any interval $[a,b]$ containing 0,

$$\int_a^b \delta(x)\,f(x)\,dx = f(0).$$

(a) Derive the Fourier series for $\delta(x)$ over $[-\pi, +\pi]$.

(b) Use (a) to suggest the value of $\delta(0)$.

(c) Let k be an arbitrary constant such that $k \in [a,b]$. What is

$$\int_a^b \delta(x-k)\,f(x)\,dx \ ?$$

(d) Use (c) to derive the Fourier series for $\delta(x-k)$ over $[-\pi, +\pi]$.

Exercise 5.29. Write a Maple procedure `fs(f(x),n)` to compute the symbolic n-term Fourier series of a function f. A sample interaction with this procedure would be

```
> fs(sin(x)*cos(x), 5);

                    1/2 sin(2 x)

> fs(x,4);

      2 sin(x) - sin(2 x) + 2/3 sin(3 x) - 1/2 sin(4 x)

> fs(x^2,5);

            2
   1/3 Pi  - 4 cos(x) + cos(2 x) - 4/9 cos(3 x) + 1/4 cos(4 x)
```

Test your Maple code against your derivations in the previous exercises.

5.4. Generalised Fourier Series and the Fourier Transform

In this section, we shall generalise the domain of the trigonometric Fourier series. We shall then use a very informal limit argument to examine how the series passes to an integral when the domain becomes infinite in extent. This integral is linked to the Fourier transform, which provides us with a continuous representation for a Fourier spectrum, a discrete version of which we have already encountered. We shall see that the Fourier transform not only confirms our intuition from previous discussions, but also can be used as a powerful analysis tool. In particular, with the use of the Fourier transform, we shall demonstrate how filtering can be performed in frequency domain.

5.4.1. Changing the Domain of a Fourier Series

The use of Fourier series can be generalised in two important ways. First, we can change the position of the interval over which the series is taken. Second, we can let the length of the interval grow or shrink arbitrarily.

Performing a Fourier series decomposition of a function over some arbitrary interval $[d,e]$ of length 2π is entirely analogous to the case over $[-\pi,\pi]$. The only difference is that the notion of even and odd functions is about the centre of the interval, namely $\dfrac{d+e}{2}$.

To address the second issue, suppose the function f of interest is periodic on an interval $I = [-L,L]$ with $L \in \mathbf{R}$, $L > 0$. We can define the Fourier series of f on I quite easily. The definition of the series itself is:

$$f(x) = \frac{a_0}{2} + \sum_{n=1}^{\infty} a_n \cos\left(\frac{n\pi x}{L}\right) + b_n \sin\left(\frac{n\pi x}{L}\right). \qquad (7)$$

The coefficients a_n, b_n are now defined as

$$a_n = \frac{1}{L}\int_{-L}^{+L} f(x)\cos\left(\frac{n\pi x}{L}\right)dx, \quad n = 0, 1, 2, \cdots .$$

$$b_n = \frac{1}{L}\int_{-L}^{+L} f(x)\sin\left(\frac{n\pi x}{L}\right)dx, \quad n = 1, 2, \cdots .$$

Notice that if we let $L = \pi$, then we get back the original form. The period of $\cos\left(\frac{n\pi x}{L}\right)$ is $\frac{2\pi}{n\pi/L}$ which simplifies to $\frac{2L}{n}$. Again, letting $L = \pi$ validates this. That means the circular frequency of $\cos\frac{n\pi x}{L}$ is $\phi_n = \frac{n}{2L}$ and that the angular frequency is $\omega_n = \frac{n\pi}{L}$.

Now, let us consider what happens as L increases. As we have seen, in many ways looking at the coefficients a_n and b_n is often even more interesting than looking at the Fourier series itself. Each a_n and b_n tell us the contribution a sinusoid of circular frequency ϕ_n and angular frequency $\omega_n = 2\pi\phi_n$ to the approximation of a function. When the size of the interval parameterised by L is allowed to grow arbitrarily large, then the function f no longer needs to be thought of as being periodic. The spacing of the discrete frequencies ϕ_n or ω_n represented by the sequence a_n becomes ever smaller, and the infinite sequence $<a_n>$ passes in the limit to the *Fourier cosine transform*:

$$f_a(\omega) \;=\; \frac{1}{\pi} \int_{-\infty}^{+\infty} f(x) \cos \omega x \, dx.$$

Correspondingly, the sequence of the b_n converge to the *Fourier sine transform*

$$f_b(\omega) \;=\; \frac{1}{\pi} \int_{-\infty}^{+\infty} f(x) \sin \omega x \, dx.$$

This argument is very informal but it provides some intuition.[7] Observe that both f_a and f_b are functions of frequency ω. Furthermore, functions like $f(x) = x$, which, as we saw above, have a Fourier series on a bounded intervals such as $[-\pi, \pi]$, do not have corresponding sine or cosine transforms on unbounded domains such as $[0, \infty]$.

Exercise 5.30. If $f(x)$ is odd, show that $f_a(\omega) = 0$. Analogously, if $f(x)$ is even, show that $f_b(\omega) = 0$.

Let us consider, for example, the function $f(x)$ that is x when $x \in [-\pi, \pi]$ and is zero outside this interval. Functions that are truncated in this manner are sometimes said to be *windowed* (which often involves a filtering operation). Observe that despite the windowing, $f(x)$ is an odd function on **R**, since for any $x \in \mathbf{R}$, $f(-x) = -f(x)$. Therefore, $f_a(\omega) = 0$, and

$$f_b(\omega) \;=\; \frac{2}{\pi} \int_0^{+\infty} x \sin \omega x \, dx$$

$$=\; 2\, \frac{\sin x\omega - x\omega\cos x\omega}{\pi\omega^2} \, \Bigg|_{x=0}^{x=\pi}$$

$$=\; 2\, \frac{\sin \pi\omega - \pi\omega\cos \pi\omega}{\pi\omega^2}.$$

A plot of this function can be found in Figure 5.22. Compare this continuous plot to the plot of the Fourier series coefficients found in Figure 5.16.

5.4.2. The Fourier Transform

When studying even and odd functions only, we can restrict ourselves to the sine and cosine transforms because one or the other will be zero, as was true for the discrete case. In general, however, to be able to represent continuous versions of a_n and b_n simultaneously requires the Fourier transform and an excursion into the complex numbers.[8] The *Fourier transform* of a function $f(x)$ is defined as

---

7. Our abbreviated discussion is in fact inspired by Marsden's treatment, pp. 395-398, *ibid.*

8. Our use of complex numbers here is minimal and should not be too difficult to follow even if the reader is unfamiliar with their general properties.

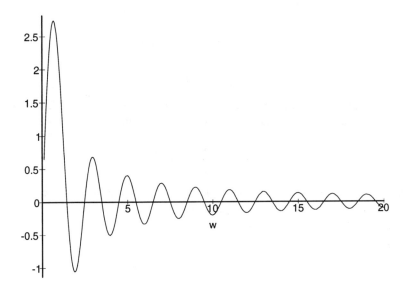

Figure 5.22. Fourier spectrum of the line $y = x$.

$$F(\omega) \;=\; \int_{-\infty}^{+\infty} f(x)\, e^{-i\omega x}\, dx,$$

and the *inverse Fourier transform* of $F(\omega)$ is

$$f(x) \;=\; \frac{1}{2\pi} \int_{-\infty}^{+\infty} F(\omega)\, e^{-i\omega x}\, d\omega,$$

where $i = \sqrt{-1}$. It is common to call F the *spectrum* of f. These integrals may appear intimidating at first, but can be made more understandable using a little intuition and using Maple.

The first thing to note is that a representation of Fourier series using complex numbers can be easily made. Two fundamental identities are that

$$e^{i\omega} \;=\; \cos\omega + i\sin\omega,$$

$$e^{-i\omega} \;=\; \cos\omega - i\sin\omega.$$

This can be verified in Maple by:

```
> evalc(exp(-I*w));
                cos(w)  -  I sin(w)
```

Clearly, the Fourier transform can be broken up into terms that look like Fourier sine and cosine transforms, modulo some scale factors that we will call k_a and k_b:

$$F(\omega) \;=\; \int_{-\infty}^{+\infty} f(x)\, e^{-i\omega x}\, dx$$

$$=\; \int_{-\infty}^{+\infty} f(x)\,(\cos \omega - i \sin \omega)\, dx$$

$$=\; k_a\, f_a(\omega) \;-\; i k_b\, f_b(\omega). \tag{8}$$

Recalling Eq. 7 for the Fourier series for a function $f(x)$ on $[-L,L]$, with some simple algebra, we can rewrite the Fourier series in complex form as

$$f(x) \;=\; \sum_{n=-\infty}^{+\infty} c_n e^{i\omega_n x},$$

where for notational clarity, we put $\omega_n = n\pi/L$, and where the complex coefficients c_n correspond to the pairs (a_n, b_n). In fact, from Eq. 8, we can express $c_n = (a_n, b_n)$ as the following integral:

$$c_n \;=\; \frac{1}{2L} \int_{-L}^{L} f(x)\, e^{-i\omega_n x}\, dx.$$

Superficially, therefore, a general Fourier series and transform have a similar form. Technically, the main difference in their use is that the series form is used to analyse *periodic* functions and the integral form is used for *aperiodic* functions. The similarity of course goes much deeper than this. We argued earlier that the coefficients (a_n, b_n) form an approximation to the Fourier spectrum or transform of a function. This is easy to verify using Maple.

As a concrete example, let us compute the Fourier transform of the box function that we considered earlier:

$$\mathrm{box}(x) \;=\; \begin{cases} 0 & \text{if } |x| > 1. \\ 1 & \text{if } |x| \le 1. \end{cases}$$

Then

$$F(\omega) \;=\; \int_{-\infty}^{+\infty} \mathrm{box}(x)\, e^{-i\omega x}\, dx$$

$$=\; \int_{-1}^{1} e^{-i\omega x}\, dx$$

$$=\; i\, \frac{e^{-i\omega x}}{\omega}\, \Bigg|_{x=-1}^{x=1}$$

$$=\; i\, \frac{e^{-i\omega}}{\omega} \;-\; i\, \frac{e^{i\omega}}{\omega}.$$

Recalling that $e^{-i\omega} = \cos\omega - i\sin\omega$, the cosine terms cancel out, and since $i^2 = -1$,

$$F(\omega) \;=\; 2\frac{\sin\omega}{\omega}$$

$$=\; 2\text{sinc}(\omega).$$

Thus the Fourier transform of a box function is a sinc, which was our prediction when we computed the Fourier series of the box function above. By symmetry, it is easy to see that the inverse Fourier transform of a box is a sinc. Notice that the result is a real-valued function. Suppose we allow the box to have width $2k$ instead. That is, let

$$\text{box}_k(x) \;=\; \begin{cases} 0 & \text{if } |x| > k. \\ 1 & \text{if } |x| \le k. \end{cases}$$

Then it is easy to see that

$$F(\omega) \;=\; 2\frac{\sin(k\omega)}{\omega}.$$

It is instructive to see the relationship between the width of the box and the resulting sinc-like function in frequency domain. Figure 5.23 depicts this relationship. Notice that there is an inverse relationship between the width of a box and the "width" of its transform. This is not surprising, given our statement above that a function cannot be both frequency-limited and space-limited, but it has strong implications for filtering, as we shall soon see.

Exercise 5.31. Using the Fourier sine and cosine transforms rather than the complex Fourier transform, go through the above derivation once again.

Exercise 5.32. Recalling Exercise 5.27, determine the Fourier transform of the Gaussian function

$$f(t) \;=\; \frac{1}{\sqrt{2\pi}\,\sigma}\,\exp\!\left(-\frac{x^2}{2\sigma^2}\right)$$

for arbitrary $\sigma > 0$. This is surprisingly easy to do by hand, but it is also an easy task for Maple. Comment on this remarkable result.

The Fourier transform has many important properties that we cannot go into here. One simple but powerful property is *linearity*. For notational convenience, let **FT**(h) denote the Fourier transform of function $h(x)$, if it exists. If $f(x)$ and $g(x)$ are two functions and $a, b \in \mathbf{R}$, then

$$\mathbf{FT}(af(x) + bg(x)) \;=\; a\mathbf{FT}(f) + b\mathbf{FT}(g),$$

if **FT**(f) and **FT**(g) exist. This property is known to Maple, which allows it (and any human) to decompose the Fourier transform of a complex function into components.

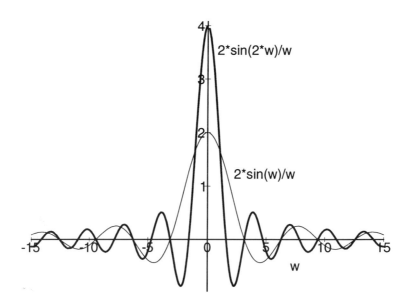

Figure 5.23. The resulting sinc functions for box widths of $2k$, $k = 1,2$. As the box width increases, the sinc function narrows and steepens.

Exercise 5.33. Extend the Fourier series package you wrote for Exercise 5.29 to arbitrary intervals. That is, write a Maple procedure `fs(f(x),n,[a,b])` to compute the n-term Fourier series of a function f on $[a,b]$.

Exercise 5.34. Use the previous exercise to compute an n-term Fourier series for $\sin x$ and $\cos x$ on $[0,\pi]$. Notice that when interpreted as periodic functions with period π, these are *not* the usual periodic sin and cos functions. Compare the Fourier series for $f(x) = x$ on $[0,\pi]$ to that on $[-\pi,\pi]$.

Exercise 5.35. Recalling Chapter 3, a *difference-of-Gaussians* filter is defined as

$$d(x;\sigma_0,\sigma_1) \;=\; f(x;\sigma_0) - g(x;\sigma_1),$$

where f and g are Gaussian distributions of standard deviation σ_0 and σ_1, respectively. Show that $\mathbf{FT}(d)$ is also a difference-of-Gaussians filter. What are the standard deviations corresponding to $\mathbf{FT}(d)$?

5.4.3. Convolution and Frequency Domain Representations

We have seen that there are mechanisms, namely the Fourier series and transform, that map both periodic and aperiodic functions $f(x)$ in spatial domain to a representation in another domain, which we have called frequency domain. That such transformations are possible is not surprising, since we have been doing them throughout this book. But we have been hinting at a particularly deep resonance between spatial and frequency representations. At the heart of this is the fundamental operation of convolution, and to what operation it transforms under the Fourier transform and its inverse. There is an entirely analogous presentation of this material that can be made using Fourier series and its related transforms, the *discrete sine, cosine,* and *Fourier* transforms. However, it is notationally more technical, so we shall deal only with continuous transforms.

Recall that the *continuous convolution* of two functions f and g, denoted $f*g$ is defined as

$$(f*g)(x) \ = \ \int_{\mathbf{R}} f(y)\,g(x-y)\,dy.$$

Also recall that $f*g = g*f$. Suppose $F(\omega) = \mathbf{FT}(f)$ and $G(\omega) = \mathbf{FT}(g)$ are the spectra (i.e., Fourier transforms) of f and g, respectively, assuming they exist. A remarkable theorem called the *convolution theorem* states that

$$\mathbf{FT}(f*g) \ = \ F\,G.$$

In other words, convolution in spatial domain is equivalent to multiplication in frequency domain. There is analogous theorem called the *modulation theorem* that expresses the duality of the converse operations:

$$\mathbf{FT}(fg) \ = \ \frac{1}{2\pi}\,(F*G).$$

We therefore have a duality: multiplication of functions in one domain is equivalent under the Fourier transform to convolution in the other.[9]

The first thing this duality allows us to do is account for why the sinc function is the only ideal low-pass filter in one dimension. Recall from Chapter 3 that filtering a signal s with a filter f is performed by convolving the signal with the filter; in other words, the filtering operation is $s*f$. Let us assume the Fourier transforms of s and f exist, and that they are denoted S and F, respectively. Then $S\,F$ is the spectrum of $s*f$. Now, for s to be a *low-pass* filter implies that the product of S and F will be such that all frequencies beyond some threshold ω_m are removed (i.e., set to zero), but that frequencies ω such that $|\omega| \leq \omega_m$ are retained, perhaps modulo a uniform scale factor. Thus F must be a *box* of half-width ω_m, and by the examples above, F, being the inverse transform of f, must be a sinc function. By the modulation theorem, frequency-domain filtering with a box is equivalent to spatial convolution with a sinc function.

To summarise, if s is a signal with frequency domain representation S, then filtering s with a sinc has the same effect in frequency domain as multiplying a box with S. Figure 5.24 illustrates the effect. In this case, our signal is represented in frequency domain and is multiplied by the Fourier transform of the

---

9. Surprisingly, these theorems are not very hard to prove. See p. 212ff of A.V. Oppenheim, A.S. Willsky, and I.T. Young, *Signals and Systems*, Prentice-Hall, Englewood Cliffs, NJ, 1983.

sinc function, which is a box in frequency domain. The result is a truncated signal occupying only the frequency band defined by the width of the box.

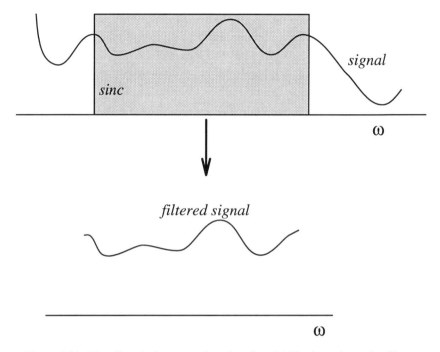

Figure 5.24. The effect, in frequency domain, of spatial filtering using a sinc filter.

Note that while we have explained the effect of filtering with a sinc from the point of view of the frequency domain, we are not required to transform to frequency domain to do filtering. However, the sinc function has an infinite domain in space, i.e., it has infinite support, but it has a finite domain in frequency. Thus it might be preferable to filter in frequency domain if it is easy to transform the signal to be filtered. This will be our next topic. A final remark to which we shall return is that for another reason, the sinc is also the ideal *reconstruction* filter. In other words, after sampling, it is the ideal basis function with which to interpolate samples to create once again a continuous function.

5.4.4. Frequency-Domain Filtering

In the previous two sections, we saw that convolution in spatial domain becomes multiplication in frequency domain, and vice versa. This suggests that it may be possible to perform filtering in frequency domain. To restate our situation, suppose a function $s(x)$ is to be filtered using filter $f(x)$, and that both s and f have Fourier transforms, $S(\omega)$ and $F(\omega)$ respectively. Then the Fourier transform of $s*f$ is SF, so to compute the result of filtering s with f, we could:

(a) Compute the Fourier transform S and F of s and f, respectively, if they exist.

(b) Compute $\overline{S} = SF$.

(c) Compute the *inverse Fourier transform* $\overline{s}(x)$ of $\overline{S}(\omega)$.

Exercise 5.36. The Gaussian distribution has another remarkable property: if f and g are two Gaussian distributions, then $f * g$ is also a Gaussian. Show that this is true. Hint: why is this exercise placed where it is?

The following Maple code summarises a frequency-domain filtering operation.

```
#
# Analytic filtering of signal s with filter f in
# frequency domain.
#
filter := proc(s,f,x)
    local S,F,SF,sf,w;
    S := evalc(fourier(s,x,w));
    F := evalc(fourier(f,x,w)):
    SF := S*F:              # NOTE: regular multiplication
    sf := evalc(invfourier(SF,w,x));
end:
```

Suppose we wish to filter the signal depicted by the thick curve in Figure 5.25 below. Actually, this "signal" is a segment of the polynomial p given by the following.

```
> p;
      (x-2) (x-1.5) (x-1) (x-.5) x (x+.5) (x+1) (x+1.5) (x+2)

> sort(expand(p),x);

       9          7             5              3
      x  - 7.50 x  + 17.0625 x  - 12.8125 x  + 2.2500 x
```

It is illustrative to use a polynomial in the filtering operations, because Maple can compute analytic Fourier transforms of polynomials. Suppose we wish to low-pass filter p. As we have seen, an ideal low-pass filter is one that is band limited in a symmetric zone about frequency $\omega = 0$. In other words, it is a box in frequency domain centred at zero; upon multiplying it in frequency domain with the function, the resulting signal only contains the low-frequency components of a function. In practice, a Gaussian (or a cubic approximation to one) is often used as a low-pass filter, despite the fact that the spectrum of a Gaussian is infinite (i.e., it is *not* band limited).[10] On the other hand, it is a smooth function that, as we have

---

10. This should be apparent if the reader has attempted the above exercises involving Gaussians.

seen, has very useful mathematical properties, and it is nicely parameterised by its
variance (or standard deviation). For a suitably-chosen variance, it acts almost
like a low-pass filter, so we shall use it in an example. We can filter p as in the
following Maple session.

```
> Digits := 5:                        # keep output size manageable
> gauss := 1/(sqrt(2*Pi)*s)*exp(-x^2/(2*s^2)):
> pg := filter(p,gauss,x):
> collect(sort(collect(pg,x),s),Pi);  # make output more readable
```

$$(3.1416 \, x^9$$
$$+ (113.10 \, s^2 - 23.562) \, x^7 + (53.605 + 1187.5 \, s^4 - 494.80 \, s^2) \, x^5$$
$$+ (536.05 \, s^2 - 40.252 + 3958.4 \, s^6 - 2474.0 \, s^4) \, x^3$$
$$+ (7.0685 - 120.76 \, s^2 - 2474.0 \, s^6 + 804.10 \, s^4 + 2968.8 \, s^8) \, x)/Pi$$

This is a complicated result, but it is very useful because it is in closed form.
We are not usually this lucky. The result is actually easy to interpret: it is a new
polynomial that is a function of both x and the standard deviation s of the
Gaussian. Observe that each coefficient in x^k is divided by π, so that the leading
coefficient is as in the original p. The subsequent coefficients are themselves
polynomials in the standard deviation s. If s is small, i.e., close to zero, then s^k
for $k > 1$ will be even smaller, and thus the constant term in each coefficient will
dominate. Upon more careful inspection, we see that this constant (after dividing
by π) is in fact the same coefficient as the corresponding one in p. On the other
hand, as s grows, it will dominate the constant term. As we see in Figure 5.25, as
s increases, it reduces the variation in p. Here, *variation* can be taken to mean
either arc length or slope. This is what we would expect of a low-pass filter: it
smooths out sharp changes in behaviour.

Exercise 5.37. Suppose we wish to design a filter that extracts from a signal the
frequencies that lie in a interval (or *band*) given by $[\omega_0, \omega_1]$. Write a Maple procedure to
generate such a *band-pass* filter. What is its spatial representation?

Exercise 5.38. Rework the example above that used Gaussians as low-pass filters to
instead use box filters in frequency domain of a given half-width.

Exercise 5.39. Do exactly as in the previous exercise, but this time, use box filters in
spatial domain of a given half-width. Apply these filters to a variety of functions, and note
the differences between them. Is a spatial box filter (i.e., a moving-average filter) a good
low-pass filter? Explain.

While low-pass filters reduce the variability in a signal, *high-pass* filters are
tuned to this variability and amplify it. Often this "tuning" is matched to the
expected class of input signals. Suppose we wish to determine where sharp
changes in the input signal occur. We might first treat the signal with a filter that

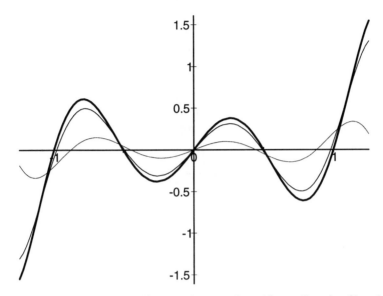

Figure 5.25. The input signal (thick curve) p together with two Gaussian-filtered signals (thinner curves). The thin curve that tracks p more closely has $s = 0.1$ and the other has $s = 0.25$.

greatly amplifies this change, and subsequently pass the result to a *thresholding* operation to pinpoint the location of the changes. Let us return to the difference of Gaussians filter that we have already seen in discrete form and introduced earlier in Exercise 5.35. In Maple, we shall define a specific filter as follows.

```
step   := Heaviside(x+1) - Heaviside(x-1):          # for later
dogn   := subs(s=0.1,gauss) - subs(s=0.2,gauss):
dog    := dogn/eval(subs(x=0,dogn))*20;
```

A plot of the function **dog** can be found in Figure 5.26; scaling this function by 20 to amplify its effect, and applying it to **step** is illustrated in Figure 5.27.

The Fourier transform and frequency-domain analysis are great mathematical and engineering triumphs, and we have only had the smallest taste of their power. These techniques were once the province of difficult pen-and-paper analysis; even now many of the integrals arising from the Fourier transform require the imagination and intuition of humans. Furthermore, the speed advantage enjoyed by approximations such as the discrete Fourier transform and the (discrete) *fast Fourier transform* will always make them the techniques of choice for time-critical applications. But the emergence of symbolic computation is helping to make analytic techniques more attractive for computational settings.

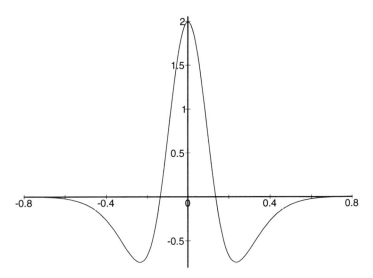

Figure 5.26. A difference of Gaussians with $\sigma_0 = 0.1$ and $\sigma_2 = 0.2$.

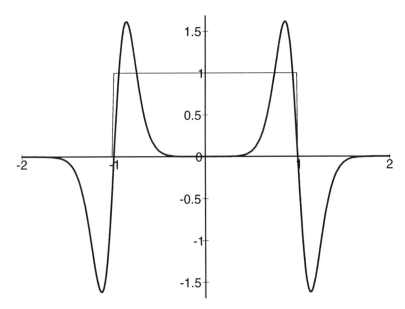

Figure 5.27. A step function and the result of filtering it with the previous difference-of-Gaussians filter (scaled by 20).

There are several other integral transforms similar to the Fourier transform that replace the *kernel* of $e^{-i\omega x}$ by another. In particular, the *Laplace transform*, which has kernel e^{-xy} is also invaluable in the study of signals, control theory, and differential equations.

5.5. The Sampling Theorem Revisited

We now have the background to restate and sketch the proof of the Sampling Theorem. Let $f(t)$ be a band-limited signal. Specifically, let the spectrum $F(\omega)$ of $f(t)$ be such that $F(\omega) = 0$ for $|\omega| > \omega_m$, for some "maximum frequency" $\omega_m > 0$. Let Δt be the spacing at which we take samples of $f(t)$. Furthermore, we define the circular *sampling rate* ω_s as

$$\omega_s = \frac{2\pi}{\Delta t}.$$

Then $f(t)$ can be uniquely represented by a sequence of samples $f(i\Delta t)$, $i \in \mathbf{Z}$ if

$$\omega_s > 2\omega_m.$$

This just states precisely what we have been saying all along: that our sampling rate must exceed twice the maximum frequency of the function.

We shall wind our way through a sketch of the proof of this theorem, because it provides us with some nice insights into Fourier series and transforms. Each step in our argument warrants more careful mathematical treatment, and a richer presentation can be found in a signal-processing textbook.[11] Nevertheless, it is a surprisingly straightforward proof. The following two exercises introduce some important properties needed later on.

Exercise 5.40. Exercise 5.28 introduced the notion of the Dirac δ-function and its sifting property. The reader who solved Exercise 3.16 in Chapter 3 will have noticed that the δ function also can be used to copy signals. Using a continuous convolution and the Dirac δ-function, show that for an arbitrary function f, and any arbitrary constant $a \in \mathbf{R}$ that

$$f(x) * \delta(x-a) = f(x-a).$$

In other words, this convolution creates a copy of f shifted by a units.

Exercise 5.41. Let $\delta_\varepsilon(x)$ denote a box of half-width ε and of area one centred at position $x = 0$. What is the height of the box? Let $f(x)$ be a function that is smooth around $[-\varepsilon, +\varepsilon]$. Then argue either pictorially or analytically that

$$f(x)\,\delta_\varepsilon(x) \approx f(0)\,\delta_\varepsilon(x).$$

(Hint: $\delta_\varepsilon(x)$ is zero outside $[-\varepsilon, +\varepsilon]$, so treat that case separately from the interior of the interval.) Now argue that as $\varepsilon \to 0$, $\delta_\varepsilon(x) \to \delta(x)$, and that

---

11. E.g., Oppenheim, Willsky and Young, *ibid*, which is the inspiration for our presentation.

$$f(x)\,\delta(x) \;\approx\; f(0)\,\delta(x).$$

This multiplication, has the effect of *sampling f* at $x = 0$. More generally, argue that

$$f(x)\,\delta(x-a) \;\approx\; f(a)\,\delta(x-a),$$

for an arbitrary $a \in \mathbf{R}$. Thus outside of an integral sign, δ works as a sampling operator.

We need an abstraction that will characterise the sampling process. Once again the δ-function comes in handy: visualise an infinite sequence of "impulses" or δ-functions, with one impulse placed at each sampling position $i\Delta T$ as in Figure 5.28. We can define this sampling train or "comb" of impulses as

$$s(t) \;=\; \sum_{i=-\infty}^{+\infty} \delta(t - i\Delta t).$$

The summation can be thought of the glue that holds a sequence of impulses together, and because $\Delta t > 0$, the impulses are spaced so that they do not overlap.

In Exercise 5.41, we saw that we could effectively sample a function f at a any desired position a by placing a δ-function at a and multiplying it with f. Therefore, multiplying f with s takes samples of f at our desired positions:

$$f_s \;=\; fs \;=\; \sum_{i=-\infty}^{+\infty} f(i\Delta t)\,\delta(t - i\Delta t).$$

This new train of "scaled" impulses is depicted in Figure 5.29.

Now we shall move to frequency domain. Suppose f and s have spectra F and S, respectively. Since f_s is a product of two functions f and s, then the modulation theorem states that its Fourier transform is a convolution:

$$\mathbf{FT}(f_s) \;=\; F_s(\omega) \;=\; \frac{1}{2\pi} F(\omega) * S(\omega).$$

For our purposes f, and therefore F, is an arbitrary function. However, we can compute the spectrum of s. It indeed turns out that the spectrum of a train of impulses of spacing Δt is *another* train of impulses in frequency domain with spacing $2\pi/\Delta t$, which we defined above to be ω_s. Formally,

$$\mathbf{FT}(s) \;=\; S(\omega) \;=\; \frac{2\pi}{\Delta t} \sum_{k=-\infty}^{+\infty} \delta(\omega - k\omega_s).$$

We can put this back into our expression for F_s:

$$F_s \;=\; \frac{1}{2\pi} F(\omega) * S(\omega)$$

$$=\; \frac{1}{2\pi} \cdot \frac{2\pi}{\Delta t}\, F(\omega) * \left(\sum_{k=-\infty}^{+\infty} \delta(\omega - k\omega_s) \right)$$

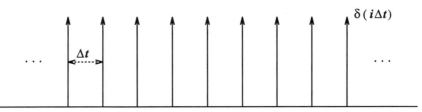

Figure 5.28. A sampling train composed of a sequence of δ-function impulses.

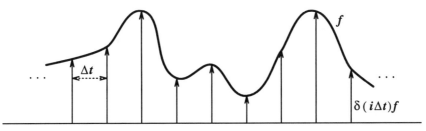

Figure 5.29. Samples of f got by multiplying f together with a sampling train.

$$= \frac{1}{\Delta t} \sum_{k=-\infty}^{+\infty} F(\omega - k\omega_s).$$

The last line comes directly from Exercise 5.40 and from the fact that convolution distributes over addition. Let us pause to interpret this result. We constructed an expression f_s for spatially sampling a function f with spectrum F. After some derivation, we have discovered that the spectrum F_s of the sampling process consists of a sum of copies of F centred at multiples of the sampling rate ω_s. Consider Figure 5.30. The upper figure depicts the spectrum F of f, which is band limited at ω_m. The middle figure is the sampling train in frequency domain. The last figure is F_s, the spectrum of the sampled function, f_s.

Recall that our goal is to establish a necessary condition for being able to reconstruct f from our samples f_s. In Figure 5.30, observe that the spacing between the copies of F is such that they do not overlap. If the spectra were to overlap, then we would be sunk: there is no hope of being able to pick out only one copy of F. This provides us with our necessary condition: the right border of one copy of F, say that with origin 0, cannot exceed the left border of the copy centred at ω_s, and analogously for the copy centred at $-\omega_s$. We have access to these values. The rightmost extent of the origin-centred copy is ω_m and the leftmost extent of the ω_s-centred copy is $\omega_s - \omega_m$. Thus

$$\omega_m < \omega_s - \omega_m.$$

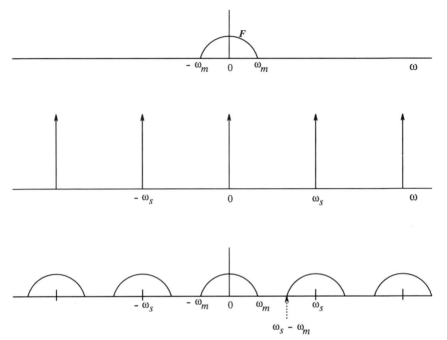

Figure 5.30. Frequency domain representation of spatial-domain sampling operation. Above: spectrum F of f. Middle: spectrum S of sampling train s.
Below: spectrum of F_s sampled signal f_s.

This implies that

$$\omega_s > 2\omega_m,$$

which establishes the bound specified in the sampling theorem, and proves the
theorem. However, our work is not yet done, because we do not yet know how to
get retrieve a single copy of F from F_s. This is the frequency-domain
characterisation of the *reconstruction* step, and it is easy to resolve: we just filter
with a box in frequency domain placed over exactly one of the copied spectra, say
the central one. Thus we need a box B of half-width ω_b no less than ω_m and no
more than $\omega_s - \omega_m$, as depicted in Figure 5.31. But we know all about our old
friend the box! The box and the sinc function are duals. With a suitably chosen
box, we have that

$$F(\omega) = F_s(\omega) B(\omega).$$

and by the convolution theorem, we can reconstruct f by convolution:

$$f(t) = f_s(t) * \text{sinc}_B(t),$$

where $\text{sinc}_B(t)$ is the inverse Fourier transform of B.

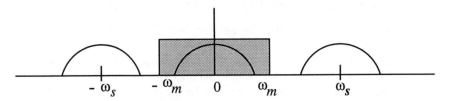

Figure 5.31. Using a box filter in frequency domain to extract one copy of the spectrum of F from F_s.

Exercise 5.42. Suppose our box B is to have width ω_b and height Δt. Then show that

$$\mathrm{sinc}_B(t) = \frac{\Delta t \omega_b}{\pi} \, \mathrm{sinc}\left(\frac{\omega_b t}{\pi}\right).$$

So now we see that not only is the sinc function the only ideal low-pass filter for one-dimensional signals, it is also the only ideal 1-D reconstruction filter. To finish our discussion, observe that if our sampling rate ω_s is less than the *Nyquist rate* $2\omega_m$, then inevitably the spectra of F in F_s will overlap, sum together and interfere with one another; unless we know something about the structure of f, there would be no way to reconstruct f exactly from the given samples. As we discussed earlier, the resulting reconstruction, even with a sinc, will invariably be a lower-frequency *alias* of f.

Appendix A: Maple Code to Compute Quadrature Rules

An extensive amount of series manipulation was required to derive the error
bound for our main quadrature rules. Maple was very helpful in deriving them.
However, a high degree of care was needed to get Maple to present the results as
we wanted them. This is by no means guaranteed to be good Maple code, but
some of manipulations may be of interest to the reader. The first script computes
the Riemann-sums Taylor series expansion.

```
terms := 5:                        # set this as desired.
#
#     Maple code to derive residual for Riemann Sums Quadrature Rule.
#
#     We have to coerce Maple to do things for us.
#
#     First, compute a series expansion about the midpoint m=(a+b)/2 of
#     the interval of interest, [a,b].
#
fx := series(f(x),x=(a+b)/2,terms):
fs := convert(fx,polynom):
fint := int(fs,x):                 # better if we handle the definite
                                   # integral on our own.

#     We have to trick Maple into doing it our way here.  We want Maple to
#     come up with a simple term-by-term expression for the integral of the
#     series.  To ensure this, we convert the polynomial sum into a list.
#     A pain in the neck?  Yes!  If someone else has a cleaner way, let me
#     know.

fl := convert(fint,list):

#     Now do a term by term evaluation of the integral over [a,b].
#     It's a little messy, so we'll do some simplifications.

fd := [seq(simplify(subs(x=b,fl[i])-subs(x=a,fl[i])),i=1..terms)]:
fd := [seq(subs(1/2*a+1/2*b=m,fd[i]),i=1..terms)]:

#     To simplify the expression, we build a list consisting of formal
#     differentiation operators [D^(i)(f) , i=0..terms-1]. We can then
#     use collect to collect up terms with respect to the differential.

i  := 'i':
DF := f:
df := f:
for i from 2 to terms do
      DF := D(DF);
      df := df, DF;
od:
i  := 'i':
df := [df]:
```

```
#     Now that we have the list, we collect and output the result back
#     as a polynomial.

print('Formal series expansion for Riemann sums');
fds := collect(fd[1],f(m)) + sum(collect(fd[i],df[i](m)),i=2..terms):
fds := subs(a-b=-dx, fds):
fds := subs(b-a=dx, fds):
fds := subs(dx=b-a, fds);
```

The next script computes the trapezoid rule residual.

```
terms := 5:                        # set this as desired.
fxm  := series(f(x),x=m,terms):
fm   := convert(fxm,polynom):            # Taylor series about m

fma  := convert(subs(x=a,fm),list):   # evaluate series at a and b
fmb  := convert(subs(x=b,fm),list):

# At this point, fma = fm + ..., and fmb = fm + ... (in list form).
# Our goal is to come up with an expression for fm in terms of fma,
# fmb, and the "...".  So we have to negate the "..." and switch
# the fma and fmb with fm.

fm2  := seq(simplify(subs(m=(a+b)/2,-(fma[i]+fmb[i])/2)), i=2..terms):

print(' ');
print('New expression for f(m)');
fm := (f(a)+f(b))/2 + sum(fm2[j], j=1..terms-1):

fm := subs(b-a=dx, fm):               # all this gets maple to write things
fm := subs(a-b=-dx, fm):           # out in terms of (b-a) rather than
fm := subs(-a+b=dx, fm):           # a mixture of (a-b) and (b-a)
fm := subs(dx=b-a, fm);

print(' ');
read ('quad.riemann'):

print(' ');
print('Let dx=(b-a) and substitute f(m) into Riemann-sums rule');
ftrap := (b-a)*fm + fds - (b-a)*f(1/2*a+1/2*b):
ftrap := subs(b-a=dx, ftrap):
ftrap := subs(a-b=-dx, ftrap):
ftrap := subs(m=(a+b)/2, ftrap):
ftrap := collect(ftrap,dx):
ftrap := subs(dx=b-a, ftrap);
```

Lastly, here is a sample session.

```
> read ('quad.trap');
```

New expression for f(m)

$$m := 1/2 \; f(a) + 1/2 \; f(b) - 1/8 \; D^{(2)}(f)(1/2 \; a + 1/2 \; b)(b - a)^2$$

$$- 1/384 \; D^{(4)}(f)(1/2 \; a + 1/2 \; b)(b - a)^4$$

Formal series expansion for Riemann sums

$$fds := (b - a) \; f(m) + 1/24 \; D^{(2)}(f)(m)(b - a)^3$$

$$+ 1/1920 \; D^{(4)}(f)(m)(b - a)^5$$

Let dx=(b-a) and substitute f(m) into Riemann-sums rule

$$ftrap := - 1/480 \; D^{(4)}(f)(1/2 \; a + 1/2 \; b)(b - a)^5$$

$$- 1/12 \; D^{(2)}(f)(1/2 \; a + 1/2 \; b)(b - a)^3$$

$$+ (1/2 \; f(a) + 1/2 \; f(b))(b - a)$$

Chapter 6
Finding the Zeros of a Function

The problem of determining the values of a parameter for which a function is zero arises in a great many applications. Many problems involving the computation of such zeros are intersection or optimisation problems, and we shall investigate some of these problems. The techniques involved are varied, and no fool-proof prescription can be given for the best technique to use for a given problem. We shall see that the representation of a function has a direct bearing on the ease of finding its zeros. Some of the techniques we shall discuss are: analytic and symbolic solutions, and numerical methods including bisection, Newton-Raphson, and secant.

6.1. Motivation: Intersection Problems

It is remarkable how many problems can be reduced to finding the set of values $x \in R^n$ for which a function $f(x)$ is zero. In numerical analysis, these problems are usually motivated by finding the zeros or the *roots* of polynomials. We will consider this problem later. As motivation, however, let us consider problems regarding the representation of, and operations with, geometric objects using implicit functions.

We saw in Chapter 1 that an *implicit* representation of a function is of the form

$$F(\mathbf{x}) = 0. \qquad (1)$$

One application of implicit functions in geometric modelling is that the set of solutions $\mathbf{x}$ for which $F(\mathbf{x}) = 0$ can be used to define geometric objects.

Example 6.1. A line $L(x,y)$ has an implicit representation

$$L(x,y) = Ax + By + C,$$

for specific $A, B, C \in \mathbf{R}$. The set

$$\{(x,y) \in \mathbf{R}^n : L(x,y) = 0\}$$

defines the points (x,y) on the line. Similarly, a plane P embedded in $\mathbf{R}^3$ is of the form

$$P(x,y,z) = Ax + By + Cz + D$$

for specific $A, B, C, D \in \mathbf{R}$. Points (x,y,z) are on the plane if $P(x,y,z) = 0$.

Example 6.2. Recalling Chapter 1, a sphere S of radius r centred at the origin has the implicit form

$$S(x,y,z) = x^2 + y^2 + z^2 - r^2. \tag{2}$$

Points (x,y,z) such that $S(x,y,z) = 0$ lie on the surface of the sphere.

Both implicit and parametric representations are convenient representations for surfaces. Implicit forms also nicely represent areas or volumes of space. For instance,

- the set of points (x,y,z) such that $S(x,y,z) \leq 0$ is a solid ball of radius r.
- the points (x,y) such that $L(x,y) < 0$ lie "below" the line L.
- the points (x,y,z) such that $P(x,y,z) > 0$ lie above the plane P, where "aboveness" is defined by the *normal* vector $\mathbf{n} = (A,B,C)$ to the plane given by $Ax + By + Cz + D = 0$.

Notions such as insideness and outsideness, which just involve inequalities when working with implicit representations, are not as easy to express using parametric representations.

We have seen that determining some geometric properties of implicit object reduce to solving for the zeros of a function. We can use geometric objects to motivate other aspects of zero finding. In computer graphics and in geometric modelling, an extremely important operation is to find the *intersection* of one object with another, namely the points two objects have in common. If A and B denote the points two objects occupy in space, then $A \cap B$ denotes their intersection. Most intersection problems can be conveniently cast as problems in finding zeros.

Let us compute the intersection between a line segment L and a plane $P(x,y,z) = Ax + By + Cz + D$. If L is embedded in the plane, then the set of solutions to $L \cap P$ is L itself. If L is embedded in a plane parallel to P, then there are no solutions; in other words, $L \cap P = \emptyset$.

Exercise 6.1. What is a simple way of determining if a line L is embedded in, or parallel to, plane P ? Write a Maple program to test for these conditions.

The final case is one in which L cuts through P at I as in Figure 6.1. In this case, $L \cap P = \{I\}$, for some $I \in \mathbf{R}^3$.

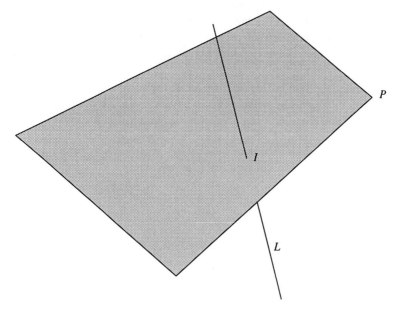

Figure 6.1. The point of intersection I of a plane P with a line L.

Line-plane intersection comes up repeatedly in computer graphics algorithms. One example is the need to *clip* a polygon with edges defined as line segments to a viewing volume whose extent is given by planes. The resulting polygon is such that it is entirely contained in the viewing volume. Another example is *ray tracing*, in which rays of light are modelled as half-infinite lines L that interact with surfaces such as planes. In any case, our goal is to compute I when it exists. A useful intersection strategy is to describe one object implicitly and the other parametrically as follows. Suppose we describe our line parametrically as

$$\mathbf{L}(t) = P_0 + t\,\mathbf{d}, \tag{3}$$

where $P_0(x_0, y_0, z_0)$ is a point on the line, $\mathbf{d} = (d_x, d_y, d_z)$ is the direction of the line, and $t \in \mathbf{R}$ is a parameter. Remember that a parametric definition is actually shorthand for three equations that must hold simultaneously, namely

$$
\begin{aligned}
x(t) &= x_0 + t\,d_x, \\
y(t) &= y_0 + t\,d_y, \\
z(t) &= z_0 + t\,d_z.
\end{aligned}
\tag{4}
$$

Our goal is to find the value of t for which the line coincides with the plane at I; that is, I must both be on the plane and on the line. Consider the following thinking. Suppose a particle Q is travelling along L. Any position of Q is characterised by a unique value of parameter t such that $Q = \mathbf{L}(t)$. If we substitute the position of Q into the plane equation, we get a value that is positive, negative or zero. We are looking for the point Q such that $P(Q) = 0$. Our strategy is therefore clear: we *substitute* the parametric equation of $\mathbf{L}(t)$ into P, and we try to solve for t. This means that we shall substitute $x(t)$ for x in P, and similarly for $y(t)$ and $z(t)$.

To keep the solution uncluttered, we recall from Appendix C of Chapter 1 the definition of a *dot* or *inner product*. If $\mathbf{a} = (a_1, a_2, a_3)$ and $\mathbf{b} = (b_1, b_2, b_3)$ are two vectors (including position vectors), then the dot product of $\mathbf{a}$ and $\mathbf{b}$, denoted $\mathbf{a} \cdot \mathbf{b}$, is defined as

$$\mathbf{a} \cdot \mathbf{b} \;=\; \sum_{i=1}^{3} a_i b_i \tag{5}$$

$$=\; a_1 b_1 + a_2 b_2 + a_3 b_3.$$

The summation form illustrates that dot products extend to vectors of arbitrary dimension. A useful property for us is that for vectors $\mathbf{a}, \mathbf{b}, \mathbf{c}$,

$$\mathbf{a} \cdot (\mathbf{b} + \mathbf{c}) \;=\; (\mathbf{a} \cdot \mathbf{b}) + (\mathbf{a} \cdot \mathbf{c}). \tag{6}$$

Furthermore, if $t \in \mathbf{R}$, then

$$\mathbf{a} \cdot (t\,\mathbf{b}) \;=\; t\,(\mathbf{a} \cdot \mathbf{b}). \tag{7}$$

Recalling that the plane equation for $P(x, y, z)$ is

$$Ax + By + Cz + D \;=\; 0,$$

by letting $\mathbf{n} = (A, B, C)$ and $\mathbf{x} = (x, y, z)$ we can rewrite P as

$$\mathbf{n} \cdot \mathbf{x} + D \;=\; 0.$$

Returning to the problem at hand, we wish to substitute the equation of the line into that of the plane and solve for the parameter t:

$$\mathbf{n} \cdot \mathbf{L}(t) + D \;=\; \mathbf{n} \cdot (P_0 + t\,\mathbf{d}) + D$$

$$=\; \mathbf{n} \cdot (P_0 + t\,\mathbf{d}) + D$$

$$=\; \mathbf{n} \cdot P_0 + t\,(\mathbf{n} \cdot \mathbf{d}) + D.$$

We are abusing notation here by supposedly taking a dot product of a vector $\mathbf{n}$ with a point P_0. We invoke our usual excuse from Chapter 2 by viewing P_0 as a position vector, and can therefore ignore the problem. But what is really happening is that we are using the dot product as a shorthand that happens to apply equally well to points and vectors.

We are almost there. Observe that the last implicit equation must equal zero, so it is easy to solve for t. In particular, if

$$\mathbf{n}\cdot P_0 + t\,\mathbf{n}\cdot\mathbf{d} + D = 0$$

then

$$t = \frac{-D-\mathbf{n}\cdot P_0}{\mathbf{n}\cdot\mathbf{d}} = \frac{-P(P_0)}{\mathbf{n}\cdot\mathbf{d}}.$$

Exercise 6.2. This solution for t is undefined if the denominator is zero. Under what conditions is $\mathbf{n}\cdot\mathbf{d}$ zero?

Thus finding the intersection of a line with a plane reduces to finding a zero of a function, and in this case we were lucky enough to have found an algebraic solution for the zero. This is not always possible, as we shall see later.

Exercise 6.3. Given two points $P_0, P_1 \in \mathbf{R}^3$, we saw in Chapter 1 that we can write a parametric equation for the line passing through these points as

$$L(t) = P_0 + t(P_1 - P_0).$$

Show that the solution in parameter t of the intersection of L with plane P is

$$t = \frac{P(P_0)}{P(P_0)-P(P_1)}.$$

A *polygon* is defined by a sequence $P_0, P_1, \cdots, P_{n-1}$ of points with the understanding that an edge joins each pair of points P_i and P_{i+1}, as well as the pair P_0 and P_{n-1}. In computer graphics, we often have to compute the intersection of a polygon with a plane by intersecting each edge with the plane. Explain why the above intersection formula for t is useful in this case.

Exercise 6.4. Suppose we wish to compute the intersection of two lines L_1 and L_2 in $\mathbf{R}^2$ and in $\mathbf{R}^3$. Formulate the intersection problem as we did above as a problem in finding the zero(s) of a function, and find an analytic solution when it exists.

In the "ray-tracing" problems of computer graphics and optics, it is particularly convenient to compute the intersection of rays with implicitly defined objects. The plane-line intersection problem that we have just considered is one example. Intersection problems involving lines and quadric surfaces such as cones and spheres are also very common. As further example, let us consider the intersection of a ray (or line) with a sphere S of radius r with centre $(0,0,0)$ as in Figure 6.2. The same approach works with spheres centred anywhere, but the notation gets messy. The symbolic manipulations are simple but a little tedious, so we shall use a symbolic computation language to help us out. The following Maple code takes us most of the way there.

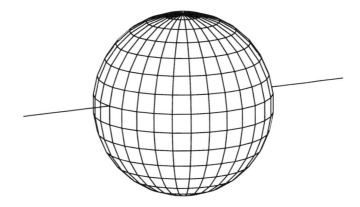

Figure 6.2. The intersection of a line with a sphere.

```
s  := x^2 + y^2 + z^2 - r^2;
xt := x0 + t*dx;
yt := y0 + t*dy;
zt := z0 + t*dz;
sp := subs(x=xt,y=yt,z=zt,s):
spc := collect(sp,t);
sols := solve(sp,t);
```

Maple tells us that the value of **spc** is

$$(dx^2 + dy^2 + dz^2)\ t^2 + (2\ x0\ dx + 2\ y0\ dy + 2\ z0\ dz)\ t$$

$$+ x0^2 + y0^2 + z0^2 - r^2$$

In slightly nicer notation, we can rewrite this as

$$spc = (d_x^2 + d_y^2 + d_z^2)\,t^2 + (2x_0 d_x + 2y_0 d_y + 2z_0 d_z)t + x_0^2 + y_0^2 + z_0^2 - r^2$$

$$= A\,t^2 + B\,t + C,$$

where $A = \mathbf{d} \cdot \mathbf{d}$, $B = 2P_0 \cdot \mathbf{d}$, and $C = P_0 \cdot P_0 - r^2$. The result is a quadratic in t, which means that there may be two solutions for t if we solve the equation $spc = 0$. This is of course expected, since a line may enter through one side of the sphere and exit through another. Our goal, therefore, is to find the *roots* of the quadratic polynomial *spc*. Fortunately, we know there is an analytic formula for the roots of a quadratic, namely that

$$t = \frac{-B \pm \sqrt{B^2 - 4AC}}{2A}.$$

The values of **sols** returned gives us symbolic expressions for t, which are too messy to reproduce here, but which the reader is invited to explore using Maple.

Exercise 6.5. Write down geometric situations in which a line L makes either zero or one intersection with sphere S. Are these geometric configurations reflected by any of the above equations?

Exercise 6.6. The above derivation gives us up to two points of intersection between a sphere and a line. In ray tracing, one is usually concerned with the *first* intersection along a ray from some point of view. How can the definition of the line L be arranged so that the first such intersection can be easily determined?

Exercise 6.7. Suppose we wish to compute the intersection of a line L and a circle of radius r and origin c in $\mathbf{R}^2$. Formulate the intersection problem as we have just done for line-sphere intersection, as a problem in finding the zero(s) of a function. Write down an analytic solution when it exists. Use Maple if you wish.

Exercise 6.8. If we have three points P_0, P_1, P_2 on a plane P, then a parametric representation for P can be written as

$$P(u,v) = P_0 + u(P_1 - P_0) + v(P_2 - P_0).$$

The three points must not all lie on a line. Why? Given this definition, use Maple to compute the intersection of two planes and the intersection of a plane with a sphere. What are the geometric objects resulting from these intersections?

The intersection problems introduced in the above discussion and exercises follow a similar pattern: to intersect two objects, find an implicit form for one and a parametric form for the other, substitute the parametric form into the indeterminates of the implicit form, and then solve for the parameters. For objects already in implicit form, such as the sphere and planes discussed above, this is an excellent strategy. In other cases, it is not. For example, using techniques beyond the scope of this book, it is possible to "implicitise" a bicubic tensor-product surface, which would allow us to compute all of the intersections between such an object and, for example, a line or another curve. This would allow us, for example, to perform analytic intersections of rays with the teapot in Chapter 2. Unfortunately, the implicit version of tensor-product surface of degree n is of degree $2n^2$. In the cubic case we would have the terrifying task of dealing with an implicit form containing more than 1300 mixed terms in the two parameters of up to degree 18.

There are other ways of performing intersections that do not require implicit representations. For example, if we have two objects defined *explicitly* as

$$y = f_1(\mathbf{x}),$$

$$y = f_2(\mathbf{x}),$$

then the points f_1 and f_2 share can be got simply by looking at the solutions to

$$f_1(\mathbf{x}) - f_2(\mathbf{x}) = 0.$$

For example, suppose our two functions are given explicitly as $y = x^3$ and $y = x^2$. We can see immediately from Figure 6.3 that the curves intersect at $x = 0$ and 1.

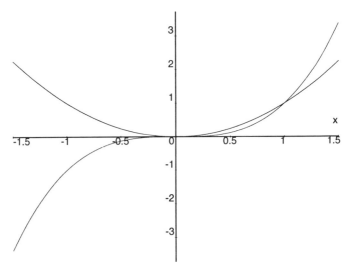

Figure 6.3.   The intersection of two polynomials, $y = x^3$ and $y = x^2$.

The set of intersections can be computed directly by computing the zeros of

$$x^3 - x^2 = x^2(x-1),$$

which leads us to the same conclusion as our figure. Since we are only interested in the zeros of the function, computing the zeros of $x^2 - x^3$ gives the same result.

6.2. Symbolic Computation of the Roots of Polynomials

Two problems have had a profound effect on the development of modern or "abstract" algebra. The first problem has been to prove that Fermat's last theorem is true.[1] This apparently has been proven recently (though there is some lingering doubt), through the efforts of thousands of mathematicians over several

---

1. Fermat's last theorem (which was really a conjecture of Fermat's, since he claimed he had a proof but did not actually produce it), states that for any integer $n \geq 3$, there are no integers x, y, z such that $x^n + y^n = z^n$.

centuries. The second problem is that of finding the roots of polynomials defined over many domains. The results of abstract algebra are deep and cannot be given justice here. Interested readers are invited to consult a modern algebra textbook. We can, however, refer to some of these results as motivation, for the "abstract" results have very concrete things to say about practical root finding.

All of our intersection problems above reduced to finding the zeros of polynomials. We derived a linear function in the case of a line-plane intersection and a quadratic function in the case of a line-sphere intersection. We also saw that there was at most one zero to the linear equation and two to the quadratic. In general, from algebra we expect n roots for a polynomial of degree n. However, there are some things we should know about these roots.

First, these roots are not always all distinct. For example, the function

$$y = x^3 \tag{8}$$

has only one distinct root, namely $x = 0$. This root is said to have *multiplicity* 3. Earlier, when we intersected the curves $y = x^2$ and $y = x^3$, we found that the solution was the polynomial $x^2(x-1)$, so the intersection point $x = 0$ is a root of multiplicity two.

Second, there is no guarantee that all the roots of a polynomial with real coefficients must be real valued. For example, a quadratic polynomial $Ax^2 + Bx + C$ will only have real roots if $B^2 - 4AC \geq 0$.

Exercise 6.9. Explain why this is true.

Third, not all polynomials can be solved analytically. There exist analytic formulae for the zeros of polynomials of degrees one to four. But a crushing blow was dealt in the early 18th Century to those believing that there would be an analytic solution for polynomials of arbitrary degree: in fact, an analytic solution for even fifth-degree polynomials cannot exist.

Throughout our discussion in this book, we have written polynomials in monomial form. We saw, however, that it was often useful to write them in other ways. One form that makes it particularly easy to pick out the roots of a polynomial is when it is *factored*: any polynomial $p(x)$ of degree n will have n possibly complex roots $r_1, r_2, \cdots, r_n$, and it can be written as

$$p(x) = (x - r_1)(x - r_2) \cdots (x - r_n), \tag{9}$$

and it may be the case that $r_i = r_j$, for some $i, j, i \neq j$. If our goal is to find all the roots of a nth-degree polynomial $p(x)$, and if we somehow have computed a root r for p, then we could factor p into

$$p(x) = (x - r)\hat{p}(x) \tag{10}$$

and $\hat{p}$ would be of degree $n - 1$. In practice, this suggests that we may be able to develop iterative algorithms to find the roots of high-degree polynomials. As we find a root, we can compute a new polynomial of one lower degree, and then proceed with it.

Symbolic computation environments usually provide a useful set of operations to deal with polynomials. To discuss these operations in any detail will require an excursion into ring theory, but a taste of the ideas is worthwhile. The mathematical structure that allows us to factor polynomials is the same as that allowing us to factor integers. The notions of prime numbers, unique factorisation, greatest common divisors, and so on, that we know from the integers have direct analogues in the polynomials. In Maple, we can factor a polynomial as follows:

```
> factor(x^3-x^2);
```
$$x^2 \; (x \; - \; 1)$$

We can get the roots of this polynomial in an obvious way:

```
> roots(x^3-x^2);
```
$$[[0, \; 2], \; [1, \; 1]]$$

This cryptic result states that the polynomial has a root of $x = 0$ with multiplicity two and a root of $x = 1$ with multiplicity one.

Two more short Maple sessions will give us a broader context. Consider doing the same operations with the polynomial $x^4 - 1$.

```
> p3 := x^4-1;
```
$$p3 \; := \; x^4 \; - \; 1$$

```
> factor(p3);
```
$$(x \; - \; 1) \; (x \; + \; 1) \; (x^2 \; + \; 1)$$

```
> roots(p3);
```
$$[[-1, \; 1], \; [1, \; 1]]$$

Maple has factored this polynomial over the rational numbers. Notice that there is still a quadratic term remaining, because it cannot be factored over the rationals. We can see immediately that there are two real roots, namely at 1 and −1. However, there may be roots over the complex numbers. We ask Maple to factor and solve the polynomial over the complex rational numbers.[2]

```
> factor(p2,I);
```
$$(x \; - \; 1) \; (x \; + \; 1) \; (x \; - \; I) \; (x \; + \; I)$$

```
> roots(p2,I);
```

---

2. Recall our definition of the rationals in Chapter 1.

$$[[I, 1], [1, 1], [-1, 1], [- I, 1]]$$

> `solve(p2);`

$$1, -1, I, - I$$

Thus we see that our roots are $x = \pm 1$ and $x = \pm i$, where $i = \sqrt{-1}$.

Now let us consider another polynomial, x^3-1. We shall factor it over the rationals and over the complex rationals.

> `p3 := x^3-1;`

$$p3 := x^3 - 1$$

> `factor(p3);`

$$(x - 1) (x^2 + x + 1)$$

> `roots(p3);`

$$[[1, 1]]$$

> `factor(p3,I);`

$$(x - 1) (x^2 + x + 1)$$

> `roots(p3,I);`

$$[[1, 1]]$$

Notice that over both, all Maple is able to do is to find one root. This is because, as it happens, the remaining roots are complex *radicals*. A radical involves an irrational number such as the square root of an integer that is not a square (e.g., $\sqrt{2}$, $\sqrt{3}$, etc.). Let us ask Maple to solve **p3**.

> `solve(p3);`

$$1, - 1/2 + 1/2\ I\ 3^{1/2}, - 1/2 - 1/2\ I\ 3^{1/2}$$

Thus our roots over the complex radicals are $x = 1$ and $x = -\frac{1}{2} \pm i\sqrt{3}$.

In passing, it is important to mention that while we have focused on the roots of polynomials in this section, symbolic systems allow us to solve for the zeros of nonpolynomials, although in this case the solution is not always possible. Consider the following examples. First we "solve" for a basic trigonometric identity.

> `solve(cos(x)^2 + sin(x)^2 = 1,x);`

$$x$$

> `solve(cos(x)^2 + sin(x)^2 = 0,x);`

$$\{null\ response\}$$

In the first case, Maple reports that the equation $\cos^2 x + \sin^2 x = 1$ for all x. Since this is true, there is no x for which the sum is 0, as is reported by Maple in the second case. The fact that a system gives no response, however, must be interpreted carefully: it may mean that no solution can exist, or it may mean that the system could not find a solution. In the case above, with some reflection we can decide which of the two outcomes apply. However, it is not always so easy to do so. The author has had several unpleasant experiences with symbolic computations running for hours, only to get a null response in return.

Next, consider a less obvious trignometric equation. The first case is a subcase of the second.

```
> solve(cos(x)^(1/2)=1/2);
```
$$arccos(1/4)$$

```
> solve(cos(x)^n = 1/2);
```
$$arccos((1/2)^{(1/n)})$$

Lastly, Maple can reason about hyperbolic equations.

```
> solve(1/x=x);
```
$$-1,\ 1$$

```
> solve(1/x=0);
```
$$\{null\ response\}$$

These are of course very simple examples of the capabilities at our disposal. Symbolic computation provides mechanisms for finding zeros that have been largely unavailable using traditional numerical techniques. They allow us to reason about functions in a way that is impossible to do numerically, and often can report the results in ways that make more sense to us.

In many cases in practice, we must use numerical methods to find the zeros of functions. Numerical methods allow us to find zeros even when we do not know what algebraic form the function (and its derivatives) has. Furthermore, in many cases, accurate numerical values are required, and the solutions must be obtained quickly, especially if they are a part of a larger application. In these cases, numerical methods should be used.

6.3. Numerical Methods for Computing Zeros

In cases where the analytic zeros cannot be computed, we are obliged to use numerical approaches. Inevitably, the approaches are approximations to the actual zero, but it is possible to get arbitrarily close to the ideal solution.

Throughout this book we have used approximations in different ways: approximations to trajectories, quadrature to approximate integrals, series representations to approximate functions and their frequency response, and so on. Numerical methods to find zeros differ from these in at least one essential way: they are *iterative*. That is, they successively approximate a root until the process converges to within the desired tolerance. In this section, we shall briefly consider several traditional approaches to the numerical computation of roots. These techniques have different applicability, depending on criteria such as the nature of the function, the spacing of its roots, and whether or not the derivative of the function is available.

6.3.1. Piecewise Approximation

Given the techniques we have already discussed in this book, it seems appropriate at first to consider a very simple strategy: to find the zeros of a function $f(x)$ on some interval $[a,b]$, first sample f at points in the interval, interpolate the samples using piecewise polynomials, and then compute the zeros of the interpolation. After all, this is essentially the theory underlying the quadrature schemes we have discussed. An algorithm might go something like this:

```
choose a sample spacing Δx on [a,b].
evaluate fᵢ = f(a + iΔx) for i = 0, 1, ···,(b−a)/Δx.
compute m curve segments pⱼ that interpolate the fᵢ.
for each curve segment pᵢ (assuming Δx is small),
     if both endpoints of pᵢ are > 0 or both are < 0,
          skip pᵢ
     if one endpoint of pᵢ is positive and the
          other is negative compute a zero
```

There are, however, several reasons why this is not usually an efficient approach:

- All the interpolation schemes discussed earlier (except the Lagrange form) presume uniformly spaced samples, say Δx, over the relevant domain. While the extension to nonuniformly spaced samples is not very difficult, it is not clear how to choose the spacing beforehand.

- If we do fix the sampling rate (i.e., knot spacing in x) globally over $[a,b]$, we do not know if we are doing too much work by taking too many samples, or missing zeros because Δx is too large.

- The quadrature schemes we discussed used polynomial interpolation as a mathematical foundation, but very efficient compound rule schemes were derived from them. At no time was an actual interpolation required. In contrast, it is not obvious how an approach based strictly on interpolation over the interval could be made efficient.

These problems have less to do with interpolation than the fact that the work–that of evaluating f and constructing interpolants–is spread evenly over the interval rather than directing the work to where it is most needed. That is precisely what iterative approaches do, as we shall now see. A simple linear or quadratic interpolant is used to push the computation in the direction of a zero, and where possible, function values are reused.

6.3.2. Bisection

Suppose we wish to compute the zero of a function $f(x)$ on an interval $[a,b]$, and suppose further that $f(a)$ and $f(b)$ are of different sign. This means that there must be at least one zero on $[a,b]$. In this case, a reliable method of finding a zero is to iteratively cut one-half of the interval away from consideration. If there is only one zero in the original interval, then our goal is to ensure that this zero is bracketed by an ever smaller interval, until either we happen upon the zero itself, or until the interval becomes small enough for us to consider the job done sufficiently well. A mathematically nice (but computationally inefficient and unstable) way of saying that two values u and v differ in sign is to say that $uv<0$. A Maple version of the bisection algorithm is:

```
#
#      Find a zero of f on interval I =[a,b].  Return result to
#      tolerance tolf.  Also stop if interval becomes < tolI.
#
bisect := proc(f, I, tolf, tolI)
    local a,b,m,fa,fb,fm,x;
    a   := I[1];
    b   := I[2];
    x   := indets(f,name)[1];        # get indeterminate
    fa  := evalf(subs(x=a,f));       # compute f(a)
    fb  := evalf(subs(x=b,f));       # compute f(b)
    m   := evalf((a+b)/2);
    while ( b-a > tolI) do           # Exercise: What if tolI is
        fm := evalf(subs(x=m,f));    #       very close to zero?
        print(m,fm);
        if (abs(fm) < tolf) then     # is fm close to 0?
            RETURN(m);
        fi;
#       Exercise: replace the following test (fa*fm < 0) by
#                 something more computer friendly.
        if (fa*fm < 0) then b := m;  # assign m to a or b
                        else a := m; #  depending on sign
        fi;
        m   := evalf((a+b)/2);
    od;
    RETURN(m);
end:
```

To compute a zero of $\cos x$ on $[-4, 1]$ to a tolerance of $0.00001 = 0.1 \times 10^{-4}$:

```
> tol := 0.00001:

> root := bisect(cos(x),[-4,1], tol, tol);

                -1.500000000, .07073720167

                -2.750000000, -.9243023786

                -2.125000000, -.5262663347

                -1.812500000, -.2393571231

                -1.656250000, -.08534970935

                -1.578125000, -.007328607602

                -1.539062500, .03172850088

                -1.558593750, .01220227396

                -1.568359375, .002436949383

                -1.573242188, -.002445858766

                                                    -5
                -1.570800782, -.4455205103*10

                    root := -1.570800782


> evalf(root + Pi/2);
                                    -5
                    -.4455*10
```

We of course know that the zero in this interval is $-\pi/2$, and we see that the desired tolerance was more than achieved. Some observations about the bisection method are worth making. First, notice that only one function evaluation is made with each iteration: f is evaluated at the midpoint of the current a and b, but one of $f(a)$ or $f(b)$ is retained from the prior computation. Second, since the interval is split in half on each iteration, we expect the error in our estimate to the root to be halved on each iteration. With more careful analysis, this can be proved.

Exercise 6.10. What will happen if there is more than one root in our interval? Run the above Maple code on several test cases.

6.3.3. The Newton-Raphson Method

The use of the derivative of a function at a point P to approximate the behaviour
of the function near P has been used extensively throughout this book. This idea
is due to Newton, and also forms the basis of an effective root-finding strategy.
Suppose we have guess x_0 for a root of a function f, and further suppose that both
f and f', the derivative of f, are available. The line that passes through $f(x_0)$
having slope $f'(x_0)$ is simply

$$L_0 = f(x_0) + f'(x_0)(x - x_0). \tag{11}$$

The reader may wish to compare this equation to our derivation of a line-drawing
algorithm in Chapter 1. The line L_0 is a local linear approximation to $f(x)$ in the
vicinity of x_0. Of course, unless f is itself a line, we do not expect L_0 to be a
particularly good approximation to f very far away from x_0, but the idea is that
the zero of L_0 may provide a good estimate of a root of f. In fact, the best for
which we can hope is that it is a better estimate than x_0, and that upon several
iterations we may converge to a root.

As above, suppose $f(x) = \cos x$, and let $x_0 = -2.5$. Then $f'(x) = -\sin x$.
Consider the following Maple code to build our first approximation to the root,
and the resulting plot in Figure 6.4.

```
x0 := -2.5;

L0 := cos(x0)-sin(x0)*(x-x0):
p  := [[x0,0],[x0,cos(x0)]]:        # vertical line to current root
plot({cos(x), L0, p},-Pi..0);
```

The line tangent to $\cos x$ at $x = -2.5$ is given by **L0** and is plotted. The zero of
any line segment can be easily calculated as follows. If

$$f(x_0) + f'(x_0)(x - x_0) = 0$$

Then

$$x = x_0 - \frac{f(x_0)}{f'(x_0)}.$$

In the specific case, then, the zero is

$$x_1 = -2.5 - \frac{\cos(-2.5)}{-\sin(-2.5)} \approx -1.16.$$

The following code and Figure 6.5 illustrate the next iteration of the method.

```
x1 := solve(L0);                    # compute intercept the lazy way
L1 := cos(x1)-sin(x1)*(x-x1):
p  := [[x1,0],[x1,cos(x1)]]:        # vertical line to current root
plot({cos(x), L1, p},-Pi..0);
```

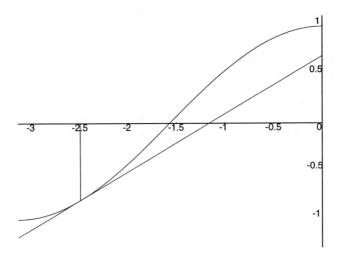

Figure 6.4. First iteration of Newton-Raphson method. The line is tangent to $y = \cos x$ at $x = -2.5$. The zero of this line at about $x = -1.16$ is our next approximation for the root.

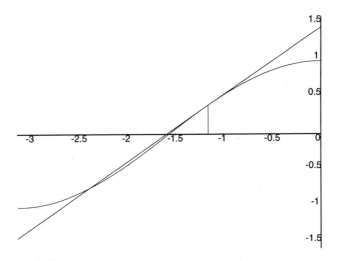

Figure 6.5. Second iteration of Newton-Raphson method. The line is tangent to $y = \cos x$ at $x = -1.16$. The zero of this line at about $x = -1.59$ is already a close approximation to our root.

With only two iterations of the method, we are already very close to a root. The overall algorithm is given by the following Maple code.

```
#
#    Find a zero of f using Newton-Raphson's method, given
#    initial guess r for a root.  Return result to tolerance tol.
#
newtonraphson := proc(f, r, tol )
    local x,               # indeterminate of f
          df,              # derivative of f
          x0, x1,          # candidates for root
          f0,              # f(x0)
          d0;              # df(x0)

    x  := indets(f,name)[1];
    df := diff(f,x);
    x0 := r;
    f0 := 2*tol;

    while  (abs(f0) > tol) do
        f0 := evalf(subs(x=x0,f));
        d0 := evalf(subs(x=x0,df));
        x1 := x0 - f0/d0;
        print(x0,x1,f0);
        x0 := x1;
    od;
    RETURN(x0);
end:
```

A sample interaction with this code follows.

```
> tol := 0.00001:
> root := newtonraphson(cos(x),-2.5, tol);

             -2.5, -1.161351872, -.8011436155

         -1.161351872, -1.595322746, .3980997644

         -1.595322746, -1.570791408, -.02452396032

                                                  -5
         -1.570791408, -1.570796327, .4918794897*10

                   root := -1.570796327

> evalf(root+Pi/2);
                        0
```

In four iterations, we have computed the root correctly to nine decimal places. This is much faster than the bisection method. Indeed when the Newton-Raphson method finds a root, it in general does it much more quickly than bisection (unless bisection happens to fall on top of a root accidentally). This method, however, has a serious difficulty, as is illustrated by the following interaction sequence. Only the final computed roots are output, and not the intermediate results.

```
> root := newtonraphson(cos(x),-0.5, tol);

                    root := -1.570796327

> root := newtonraphson(cos(x),-0.4, tol);

                    root := -4.712388980

> root := newtonraphson(cos(x),-0.1, tol);

                    root := -10.99557429


> root := newtonraphson(cos(x),-0.05, tol);

                    root := -20.42035225

> root := newtonraphson(cos(x),0.00, tol);

              Error, (in newtonraphson) division by zero
```

We clearly have a serious problem. If we change the initial guess slightly for a root, we can get a different root as a result. Even more catastrophic is that for one choice of a root, the routine fails on a division by zero. The ultimate humiliation is that for some functions, the routine never returns! What is happening? Let us examine what happens if the initial guess is close to a maximum or minimum of a function.

Figure 6.6 depicts the results if our initial guesses were $x_0 = -0.2, -0.05$, and -0.00001. The line whose intercept would be the next root is increasingly horizontal as x nears zero. When $x_0 = -0.2$, the next guess for a root is shot well beyond the root closest to $x_0 = -0.2$. Indeed it is put in the vicinity of the second root at $3\pi/2$. The pictorial evidence would suggest that since we are not near a maximum of $\cos x$, we should quickly converge to this value. That is indeed what happens.

Similarly, when $x_0 = -0.5$, the next guess for a root is pushed out somewhere near the tenth root of $\cos x$. However, we can see that when x is very close to zero, the intercept of the resulting line is a poor guess for *any* root.

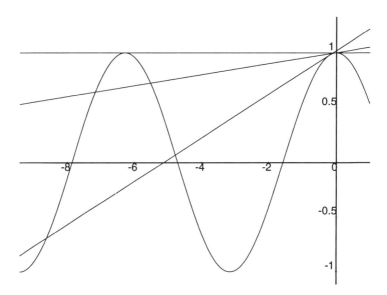

Figure 6.6. The effect, using the Newton-Raphson method, of an initial guess that
is close to a maximum or minimum of the function. The three lines
result from taking $x_0 = -0.2, -0.05, -0.000001$. The resulting lines are
more horizontal as x gets closer to a maximum of $\cos x$ at 0.

The advantage of the Newton-Raphson method is that when the derivative is
not too small, then it finds a root very quickly. Its disadvantages are: that the root
it finds may not be the obvious one, that it is both unstable and may converge
slowly when the derivative is small, and that it requires both the function and its
derivative to be available. In the cases above, the derivative was readily available
and was as cheap to compute as the function itself. In many cases, however, we
may not even have an analytic form for f, let alone its derivative. Moreover,
derivatives are often more numerically expensive and less stable to compute than
the function. We shall address this issue next.

Exercise 6.11. Give examples of functions whose derivatives are both more expensive and
less stable (i.e., prone to numerical inaccuracies) to compute. Hint: consider rational
functions of the form $f(x)/g(x)$, where $f(x)$ and $g(x)$ involve functions such as square-root
and exp.

6.3.4. The Secant Method

Because derivatives may not be available or may be difficult to compute, we can instead approximate the derivative to a function f. The easiest way to do this is to use two samples of f, and estimate f' by the slope of the line connecting the two samples. Thus at the expense of some extra function evaluations, we can come up with a coarse approximation for a derivative. As we saw with the Newton-Raphson method, we computed a sequence $x_0, x_1, \cdots, x_n$ of potential roots, with x_n hopefully being within a user-specified tolerance. Thus if we begin with values $(x_0, f(x_0))$ and $(x_1, f(x_1))$, then we can use these values to compute x_2. We then use $(x_1, f(x_1))$ and $(x_2, f(x_2))$ to compute x_3 and so on. More formally, if $(x_{i-1}, f(x_{i-1}))$ and $(x_i, f(x_i))$ are the two most recent points in our iteration, then an approximation for the derivative of f at x_i is

$$f'(x_i) \approx \frac{f(x_i) - f(x_{i-1})}{x_i - x_{i-1}}$$

Thus we can rewrite our Newton-Raphson line, Eq. 11, as follows.

$$L_i = f(x_i) + (x - x_i) f'(x_i)$$
$$\approx f(x_i) + (x - x_i) \frac{f(x_i) - f(x_{i-1})}{x_i - x_{i-1}}.$$

The zero of this line segment is our next estimate for the root of f, x_{i+1}. When

$$L_i = 0,$$

we have that

$$x = x_i - f(x_i) \frac{x_i - x_{i-1}}{f(x_i) - f(x_{i-1})}.$$

We set x_{i+1} to this value and iterate as we did for the Newton-Raphson method. This technique is called the *secant method*. As would be expected, the convergence rate of this method–namely the quality of the solution with respect to the number of function evaluations–is slightly below the Newton-Raphson method. However, in many cases, the cost of computing an exact derivative can be much higher than the original function, so that in this case the small number of extra function evaluations and arithmetic operations may turn out to be an economy.

Exercise 6.12. Modify the Newton-Raphson Maple code to a routine called **secant** that implements the secant method. Use this routine for the same functions as above and comment on the behaviour of the secant method as compared with the Newton-Raphson method. Compare the number of iterations needed for each method on a variety of functions. Also, use Maple's **time** procedure to get the computation time for each routine. While this is not as indicative as a C implementation, comment on whether this time measure is consistent with the number-of-iterations measure.

Exercise 6.13. Given the root-finding techniques discussed in this chapter, suggest a hybrid algorithm that is usually as fast as the secant or Newton-Raphson methods, but that does not misbehave near a local optimum of the function. A great many numerical techniques are in fact hybrids of various algorithms. Such algorithms characterise possibly problematic behaviour and invoke the method best suited to that behaviour.

Index